AQA Mathematics for GCSE

Exclusively endorsed and approved by AQA

Series Editor
Paul Metcalf

Series Advisor
David Hodgson

Lead Author
Steven Lomax

June Haighton
Anne Haworth
Janice Johns
Andrew Manning
Kathryn Scott
Chris Sherrington
Margaret Thornton
Mark Willis

HIGHER

Module 1

Nelson Thornes
a Wolters Kluwer business

Published in 2006 by:
Nelson Thornes Ltd
Delta Place
27 Bath Road
CHELTENHAM
GL53 7TH
United Kingdom

06 07 08 09 10 / 10 9 8 7 6 5 4 3 2 1

A catalogue record for this book is available from the British Library.

ISBN 0 7487 9755 6

Cover photograph: Kingfisher by Paulo De Oliveira/OSF/Photolibrary
Illustrations by Roger Penwill
Page make-up by MCS Publishing Services Ltd, Salisbury, Wiltshire

Printed in Great Britain by Scotprint

Acknowledgements

The authors and publishers wish to thank the following for their contribution:
David Bowles for providing the Assess questions
David Hodgson for reviewing draft manuscripts

Thank you to the following schools:
Little Heath School, Reading
The Kingswinford School, Dudley
Thorne Grammar School, Doncaster

The publishers thank the following for permission to reproduce copyright material:

Explore photos
Diver – Corel 55 (NT); Astronaut – Digital Vision 6 (NT);
Mountain climber – Digital Vision XA (NT); Desert explorer – Martin Harvey/Alamy.

Archery – AAS; Plant – Stockbyte 29 (NT); Students in exam – Digital Stock 10 (NT);
Beach – Corel 777 (NT); Factory worker – Digital Stock 7 (NT); House sale – Photodisc 76 (NT);
Vitruvian man – Corel 481 (NT); Crufts dog show – Homer Sykes/Alamy;
Shopping centre – Corel 641 (NT); Telephone survey – Photodisc 55 (NT);
Students sitting exam – Image 100 37 (NT); Cricket – Corel 449 (NT).

The publishers have made every effort to contact copyright holders but apologise if any have been overlooked.

Contents

Introduction

This book has been written by teachers and examiners who not only want you to get the best grade you can in your GCSE exam but also to enjoy maths.

Each chapter has the following stages:

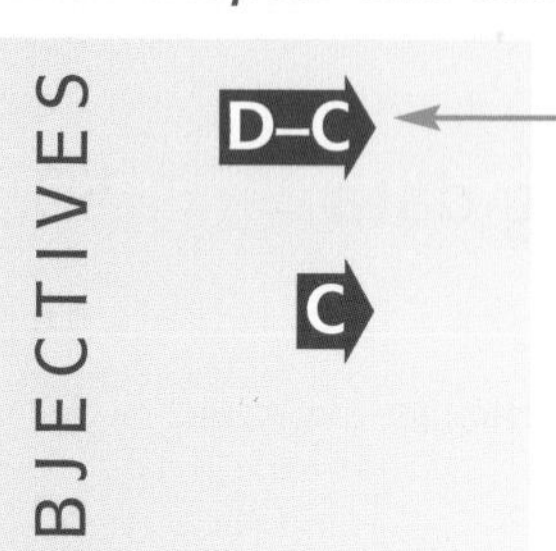

The objectives at the start of the chapter give you an idea of what you need to do to get each grade. Remember that the examiners expect you to perform well at the lower grade questions on the exam paper in order to get the higher grades. So, even if you are aiming for an A grade you will still need to do well on the D grade questions on the exam paper.*

Learn 1 — *Key information and examples to show you how to do each topic. There are several Learn sections in each chapter.*

Apply 1 — *Questions that allow you to practise what you have just learned.*

(calculator icon) — *Means that these questions should be attempted with a calculator.*

(crossed-out calculator icon) — *Means that these questions are practice for the non-calculator paper in the exam and should be attempted without a calculator.*

Get Real! — *These questions show how the maths in this topic can be used to solve real-life problems.*

<u>1</u> — *Underlined questions are harder questions.*

Explore — *Open-ended questions to extend what you have just learned. These are good practice for your coursework task.*

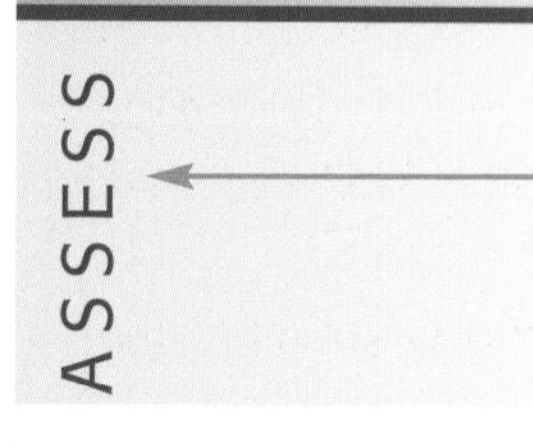

End of chapter questions written by an examiner.

Some chapters feature additional questions taken from real past papers to further your understanding.

1 Statistical measures

OBJECTIVES

D **Examiners would normally expect students who get a D grade to be able to:**

Calculate the mean for a frequency distribution

Find the modal class for grouped data

C **Examiners would normally expect students who get a C grade also to be able to:**

Find the mean for grouped data

Find the median class for grouped data

What you should already know ...

- Multiplication and division of a set of numbers without the use of a calculator
- Find the mean, median, mode and range for a set of numbers
- Compare the mean and range of two distributions
- Understand the inequality signs <, >, ⩽ and ⩾

VOCABULARY

Frequency table or **frequency distribution** – a table showing how many times each quantity occurs, for example:

Number in family	2	3	4	5	6	7	8
Frequency	2	3	8	4	2	0	1

Average – a single value that is used to represent a set of data

Mean – found by calculating

$$\frac{\text{the total of all the values}}{\text{the number of values}}$$

Mode – the value that occurs most often

Modal class – the class with the highest frequency

Median – the middle value when all the values have been arranged in order of size; for an even set of numbers, the median is the mean of the two middle values

Range – a measure of spread found by calculating the difference between the largest and smallest values in the data, for example, the range of 1, 2, 3, 4, 5 is 5 – 1 = 4

Data – information that has been collected

Discrete data – data that can only be counted and take certain values, for example, number of cars (you can have 3 cars or 4 cars but nothing in between, so $3\frac{1}{2}$ cars is not possible)

Continuous data – data that can be measured and take any value; length, weight and temperature are all examples of continuous data

Grouped data – data that has been grouped into specific intervals

Learn 1 Mean, mode, median and range for a frequency table

Example:

From this table work out the mean, mode, median and range of the number of goals scored.

Goals scored (x)	Frequency (f)
0	3
1	6
2	5
3	3
4	2
5	1

Completing the table:

Goals scored (x)	Frequency (f)	Frequency × goals scored (fx)
0	3	$0 \times 3 = 0$
1	6	$1 \times 6 = 6$
2	5	$2 \times 5 = 10$
3	3	$3 \times 3 = 9$
4	2	$4 \times 2 = 8$
5	1	$5 \times 1 = 5$
Total	$\Sigma f = 3 + 6 + 5 + 3 + 2 + 1 = 20$	$\Sigma fx = 0 + 6 + 10 + 9 + 8 + 5 = 38$

Σ means 'the sum of' so Σf means the sum of the frequencies and Σfx means the sum of the (frequency × goals scored)

The mean $= \frac{\Sigma fx}{\Sigma f} = \frac{38}{20} = 1.9$ goals

The mode is the value that occurs the most often.

Mode = 1 goal — In a frequency table, this is the value with the highest frequency. The highest frequency is 6 so the mode is 1

The median is the middle value when all the values have been arranged **in order of size.**

The middle value is the $\frac{\Sigma f + 1}{2}$ th value $= \frac{20 + 1}{2}$ th value = 10.5th value.

So the median lies between the 10th and 11th value.

To find the 10th and 11th terms do a quick running total of the frequencies.

Goals scored (x)	Frequency (f)	Running total
0	3	3
1	6	$3 + 6 = 9$
2	5	$9 + 5 = 14$
3	3	
4	2	
5	1	

The 10th and 11th terms are in this class so the median = 2 goals

Range
Largest number = 5
Smallest number = 0

The range is the difference between the largest and smallest numbers = 5 – 0 = 5

Apply 1

1 The shoe sizes of ten people are shown in the table.

Shoe size (x)	Frequency (f)	Frequency × shoe size (fx)
3	3	
4	3	
5	4	

a Copy and complete the table.

b Find the mean shoe size.

2 A dice is thrown 100 times. The scores are shown in the table.

Score (x)	Frequency (f)	Frequency × score (fx)
1	18	
2	19	
3	16	
4	12	
5	15	
6	20	

Copy and complete the table and find the mean score.

3 Ten people were asked to give the ages of their cars. Their answers are shown in the table.

Age of car (x)	Frequency (f)
1	2
2	3
3	4
4	1

Tom says that the mean age of the cars is 6 years.

a Find the mean.

b What do you think Tom did wrong?

4 **Get Real!**

Andrew has five people in his family. He wondered how many people there were in his friends' families. He asked 20 of his friends and put his results in a table.

Number in family	2	3	4	5	6	7	8
Frequency	2	3	8	4	2	0	1

From the data calculate:

a the median

b the mode

c the mean

d the range.

e Which average is the most appropriate to use? Why?

5 Get Real!

A group of 30 teenagers were learning archery and were allowed ten shots each at the target. The instructor counted the number of times they hit the target and recorded the following results.

2 6 2 0 2 3 8 7 2 5 6 1 7 4 8
8 6 1 0 3 2 2 1 3 8 3 2 6 7 5

a Copy the table and use the figures to complete it.

Number of hits	0	1	2	3	4	5	6	7	8
Frequency									

b Use the table to calculate:

i the mode **ii** the median **iii** the mean **iv** the range.

c Which average is the most appropriate to use with this data? Why?

6 Get Real!

Rachel is investigating the number of letters in words in her reading book. Her results are shown in the following table.

Number of letters in a word	Frequency
1	5
2	6
3	10
4	4
5	7
6	8
7	4
8	2

She says, 'The mode is 10, the median is 4 and the mean is 4.2'.
Is she right? Give a reason for your answer.
Which average do you think best represents the data? Why?

7 Get Real!

Mr Farrington, a technology teacher, has estimated that the average length of a box of bolts is 60 mm, to the nearest millimetre.
The class measured the lengths of 20 boxes and recorded their results in a table.

Length (mm)	Frequency	Frequency × length
58	3	
59	4	
60	2	
61	2	
62	8	
63	1	
Total		

Copy and complete the table.
Work out the mean length.
Was Mr Farrington correct in his estimation?

8 Get Real!

These marks were obtained by a class of 28 students in a science test. The maximum mark possible was 25.

20 17 21 15 16 15 12 14 15 19 21 17 20 22
16 19 21 13 22 15 14 18 20 13 18 16 15 19

By drawing a frequency table:

a Calculate the mean and work out the median and mode.

b Work out the range of the marks.

c What would you give as the pass mark? Explain your answer.

9 The mean of five integers is 4 and the median is 3. Find three different groups of numbers that fit this statement.

10 Get Real!

In five schools, the heights of a sample of the students were recorded and the mean calculated. The results are shown in the table.

School	Sample size	Mean height
A	150	1800 mm
B	200	1770 mm
C	400	1810 mm
D	350	1830 mm
E	400	1810 mm

a What is the mean height of the complete sample over all the schools?

b How many schools show mean heights below the overall mean?

Explore

A DIY shop sells a selection of large letters that can be used to design name boards for houses

The shopkeeper wants to buy 1000 letters but realises that some letters will be more popular than other letters

For example

needs two Es, two Os, two Ts and one of each of the other letters

How many of each letter should the shopkeeper buy?

Investigate further

Explore

- Draw a straight line on a sheet of paper
- Place a ruler some distance away from the line
- Let several people look at the line and the ruler
- Ask each one to estimate the length of the line
- Put your results into a frequency table and calculate the mean
- How does this mean compare with the actual length of the line?

Investigate further

Learn 2 Mean, mode and median for a grouped frequency table

Example:

The weights of 50 potatoes are measured to the nearest gram and shown in the table below.

From this table work out the mean, modal class and the class containing the median.

Weight in grams	Frequency
75–79	3
80–84	3
85–89	3
90–94	10
95–99	7
100–104	7
105–109	5
110–114	4
115–119	2
120–124	4
125–129	1
130–134	1

The weight of potatoes is continuous

The class 90–94 means any weight between 89.5 and 94.5

The lower bound is 89.5 and the upper bound is 94.5

The midpoint of the class is

$$\frac{89.5 + 94.5}{2} = 92$$

Classes can be written in different ways:

for example 124.5 up to 129.5
$124.5 \leqslant x < 129.5$
$124.5 < x \leqslant 129.5$
etc

Completing the table:

Weight in grams	Frequency (f)	Midpoint (x)	Frequency × midpoint (fx)
75–79	3	77	$77 \times 3 = 231$
80–84	3	82	$82 \times 3 = 246$
85–89	3	87	$87 \times 3 = 261$
90–94	10	92	$92 \times 10 = 920$
95–99	7	97	$97 \times 7 = 679$
100–104	7	102	$102 \times 7 = 714$
105–109	5	107	$107 \times 5 = 535$
110–114	4	112	$112 \times 4 = 448$
115–119	2	117	$117 \times 2 = 234$
120–124	4	122	$122 \times 4 = 488$
125–129	1	127	$127 \times 1 = 127$
130–134	1	132	$132 \times 1 = 132$
Totals	$\Sigma f = 50$		$\Sigma fx = 5015$

Start by finding the midpoint of each class and then continue as a frequency table

$\Sigma fx = 231 + 246 + 261 + 920 + 679 + 714 + 535 + 448 + 234 + 488 + 127 + 132$

The mean $= \dfrac{\Sigma fx}{\Sigma f} = \dfrac{5015}{50} = 100.3$ g

Remember that this is only an estimate as you have used the midpoints

The modal class is the class that occurs the most often.

Modal class = 90–94 g

In a frequency table, this is the class with the highest frequency

The median is the middle value when all the values have been arranged **in order of size.**

The middle value is the $\dfrac{\Sigma f + 1}{2}$ th value $= \dfrac{50 + 1}{2}$ th value = 25.5th value.

So the median lies between the 25th and 26th value.

To find the 25th and 26th terms do a quick running total of the frequencies.

Weight in grams	Frequency (f)	Running total
75–79	3	3
80–84	3	$3 + 3 = 9$
85–89	3	$6 + 3 = 9$
90–94	10	$9 + 10 = 19$
95–99	7	$19 + 7 = 26$
100–104	7	
105–109	5	
110–114	4	
115–119	2	
120–124	4	
125–129	1	
130–134	1	

The 25th and 26th values are in this class so the class containing the median = 95–99 g

Apply 2

1 Get Real!

The table shows the wages of 40 staff in a small company.

Wages (£)	Frequency
$50 \leqslant x < 100$	5
$100 \leqslant x < 150$	13
$150 \leqslant x < 200$	11
$200 \leqslant x < 250$	9
$250 \leqslant x < 300$	0
$300 \leqslant x < 350$	2

Find:

a the modal class

b the class that contains the median

c an estimate of the mean.

2 Get Real!

The scores obtained in a survey of reading ability are given in this table.

Reading scores (x)	Frequency
$0 \leqslant x < 5$	15
$5 \leqslant x < 10$	60
$10 \leqslant x < 15$	125
$15 \leqslant x < 20$	260
$20 \leqslant x < 25$	250
$25 \leqslant x < 30$	200
$30 \leqslant x < 35$	90

a What is the modal class?

b Calculate an estimated mean reading score.

3 Get Real!

The lengths, to the nearest millimetre, of a sample of a certain type of plant are given below.

51 51 58 54 59 60 52 52 55 49 51 53
55 60 58 57 51 57 56 50 53 58 59 57

a Calculate the mean length.

b Calculate an estimate of the mean length by grouping the data in class intervals of 47–49, 50–52, etc.

c Comment on your findings. What do you notice?

4 For each of these sets of data, work out the:

a mean **b** modal class **c** class containing the median.

i

Mark	Frequency
21 up to 31	1
31 up to 41	1
41 up to 51	3
51 up to 61	9
61 up to 71	8
71 up to 81	6
81 up to 91	2

ii

Daily takings ($)	Frequency
480–499	2
500–519	3
520–539	5
540–559	7
560–579	11
580–599	13
600–619	6
620–639	4
640–659	0
660–679	1

5 The table shows the weights of ten letters.

Weight (grams)	$0 \leqslant x < 20$	$20 \leqslant x < 40$	$40 \leqslant x < 60$	$60 \leqslant x < 80$	$80 \leqslant x < 100$
Number of letters	2	3	2	2	1

Calculate an estimate of the mean weight of a letter.

6 A survey was made of the amount of money spent at a supermarket by 20 shoppers. The table shows the results.

Amount spent, A (£)	$0 \leqslant A < 20$	$20 \leqslant A < 40$	$40 \leqslant A < 60$	$60 \leqslant A < 80$
Number of shoppers	1	7	8	4

Calculate an estimate of the mean amount of money spent by these shoppers.

Explore

- Sarah has collected some data from students
- She has found that the 'average' student is male, has brown eyes and hair, and is 165 cm tall
- Is this true for the students in your class?
- What else can you say about students in your class?

Investigate further

Statistical measures

ASSESS

The following exercise tests your understanding of this chapter, with the questions appearing in order of increasing difficulty.

1 The table shows the number of trainers sold in one day in a sports shop.

Size	5	$5\frac{1}{2}$	6	$6\frac{1}{2}$	7
Frequency	10	15	9	3	1

Find the mode, median, mean and range of this data.

2 J A Williams has a dental surgery. The information below shows the waiting times for patients during one day.
Find the class intervals that contain the mode and median and calculate an estimate of the mean waiting time at J A Williams' surgery.

Waiting time, x (minutes)	$0 \leqslant x < 3$	$3 \leqslant x < 6$	$6 \leqslant x < 9$	$9 \leqslant x < 12$	$12 \leqslant x < 15$	$15 \leqslant x < 18$
Number of patients	9	15	12	8	4	2

3 Debbie asks some of the students in her class how many brothers and sisters they have. She puts the information in a table.

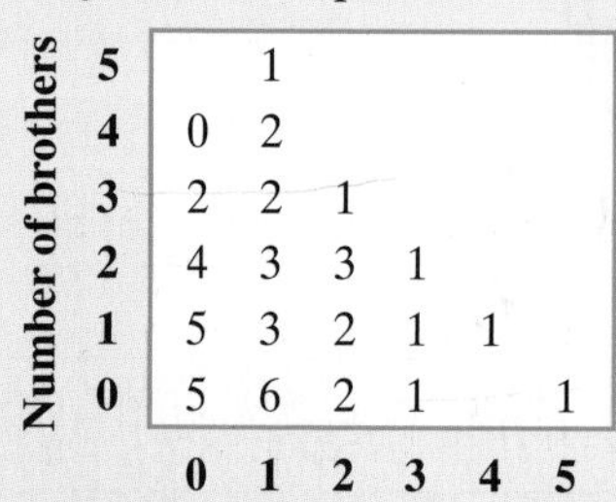

Number of brothers \ Number of sisters	0	1	2	3	4	5
5		1				
4	0	2				
3	2	2	1			
2	4	3	3	1		
1	5	3	2	1	1	
0	5	6	2	1		1

a How many students have no sisters?

b How many students have only one brother?

c How many students have equal numbers of brothers and sisters?

d How many students did Debbie survey altogether?

e What is the modal number of sisters?

f Calculate the mean number of brothers.

4 The information below shows the speeds of 60 white vans passing a speed camera. Find the class intervals that contain the mode and median and calculate an estimate of the mean.

Speed, x (mph)	$30 \leqslant x < 40$	$40 \leqslant x < 50$	$50 \leqslant x < 60$	$60 \leqslant x < 70$	$70 \leqslant x < 80$	$80 \leqslant x < 90$	$90 \leqslant x < 100$
Frequency	2	10	18	16	11	2	1

5 An office manager monitored the time members of staff took on 'private' telephone calls during working hours. Calculate an estimate of the mean length of the telephone calls, giving your answer to the nearest minute.

Time (nearest min)	3–5	6–8	9–12	13–16	17–20	21–25	26–30	31–40
Frequency	67	43	28	13	7	4	3	2

2 Representing data

OBJECTIVES

D

Examiners would normally expect students who get a D grade to be able to:

Construct a stem-and-leaf diagram (ordered)

Construct a frequency diagram

Interpret a time series graph

B

Examiners would normally expect students who get a B grade also to be able to:

Construct and interpret a cumulative frequency diagram

Use a cumulative frequency diagram to estimate the median and interquartile range

Construct and interpret a box plot

Compare two sets of data using box plots

Construct a time series and plot the moving averages

Use the trend line to estimate other values

A

Examiners would normally expect students who get an A grade also to be able to:

Construct and interpret a histogram with unequal class intervals

What you should already know ...

- Measures of average including mean, median and mode
- Accurate use of ruler and protractor
- Construct and interpret a pictogram, bar chart and pie chart
- Interpret a stem-and-leaf diagram
- Interpret a time series graph

VOCABULARY

Frequency diagram – a frequency diagram is similar to a bar chart except that it is used for continuous data. In this case, there are usually no gaps between the bars

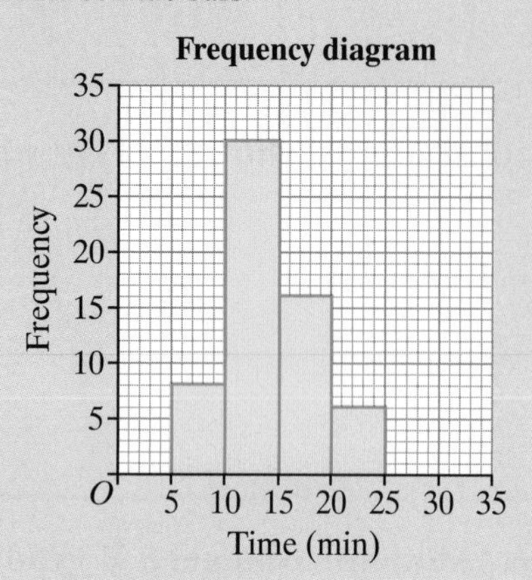

Stem-and-leaf diagram – a way of arranging data using a key to explain the 'stem' and 'leaf' so that 3|4 represents 34

Number of minutes to complete a task

Stem (tens)	Leaf (units)
1	6 8 1 9 7
2	7 8 2 7 7 2 9
3	4 1 6

Key: 3|4 represents 34 minutes

Ordered stem-and-leaf diagram – a stem-and-leaf diagram where the data is placed in order

Number of minutes to complete a task

Stem (tens)	Leaf (units)
1	1 6 7 8 9
2	2 2 7 7 7 8 9
3	1 4 6

Key: 3|4 represents 34 minutes

Back-to-back stem-and-leaf diagram – a stem-and-leaf diagram used to represent two sets of data

Number of minutes to complete a task

Leaf (units) Girls	Stem (tens)	Leaf (units) Boys
7 7 6 5 4 2 2	1	1 6 7 8 9
7 6 4 3 2 1	2	2 2 7 7 7 8 9
7 0	3	1 4 6

Key: 3|2 represents 23 minutes

Key: 3|4 represents 34 minutes

Line graph – a line graph is a series of points joined with straight lines

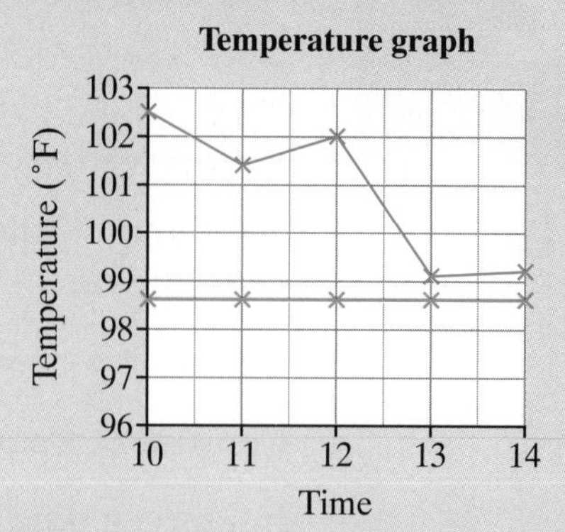

Time series – a graph of data recorded at regular intervals

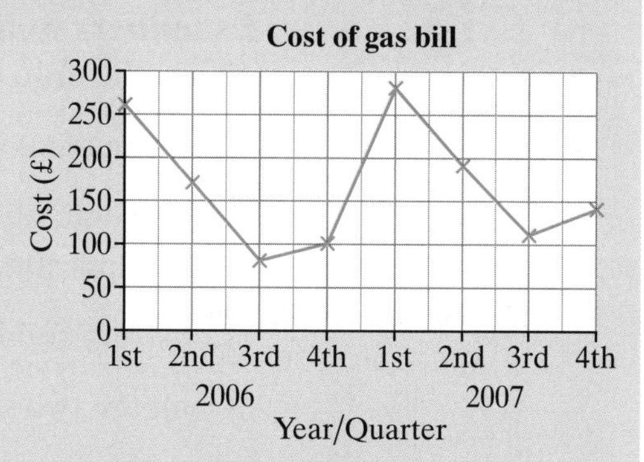

Moving average – used to smooth out the fluctuations in a time series, for example, a four-point moving average is found by averaging successive groups of four readings

The four-point moving averages can be plotted on the graph as shown

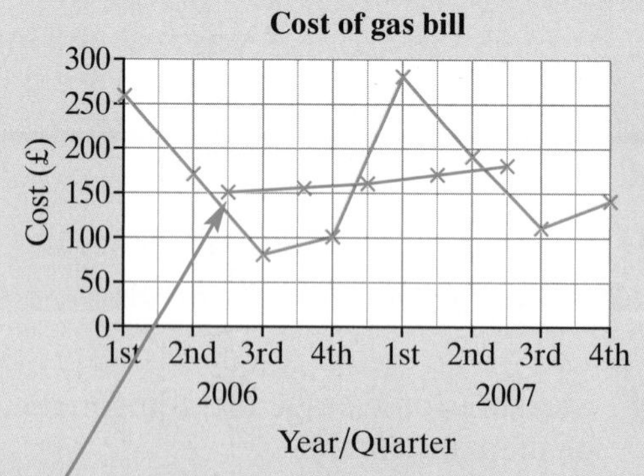

The first four-point moving average is plotted in the 'middle' of the first four points, and so on

Cumulative frequency diagram – a cumulative frequency diagram can be used to find an estimate for the mean and quartiles of a set of data; find the cumulative frequency by adding the frequencies in turn to give a 'running total'

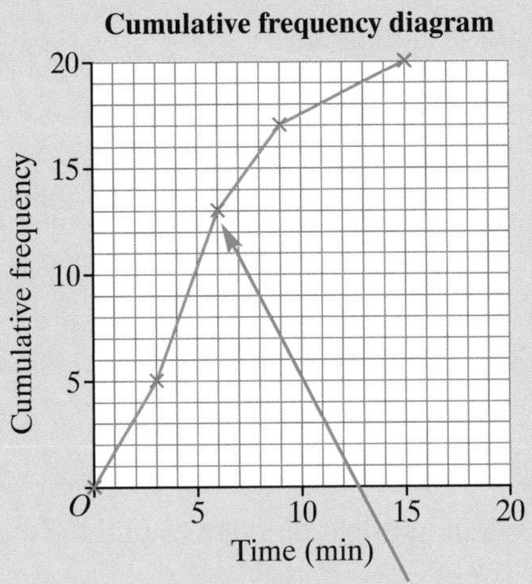

The cumulative frequencies are plotted at the end of the interval to which they relate

Median – the middle value when all the values have been arranged in order of size; for an even set of numbers, the median is the mean of the two middle values

Lower quartile – the value 25% of the way through the data

Upper quartile – the value 75% of the way through the data

Interquartile range – the difference between the upper quartile and the lower quartile

$$\text{Interquartile range} = \begin{matrix}\text{upper}\\\text{quartile}\end{matrix} - \begin{matrix}\text{lower}\\\text{quartile}\end{matrix}$$

Box plot or **box and whisker plot** – used to show how the data is distributed

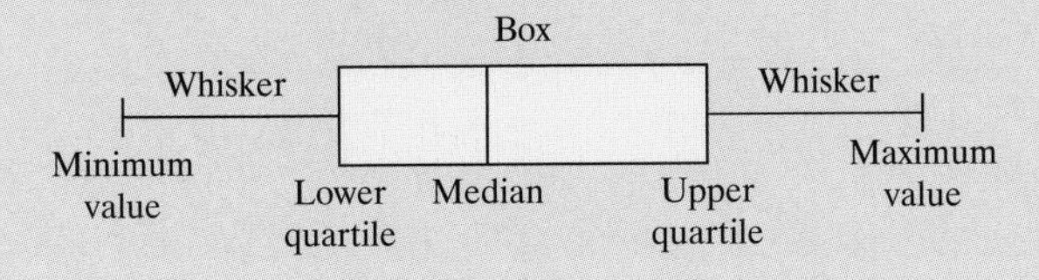

Histogram – a histogram is similar to a bar chart except that the *area* of the bar represents the frequency

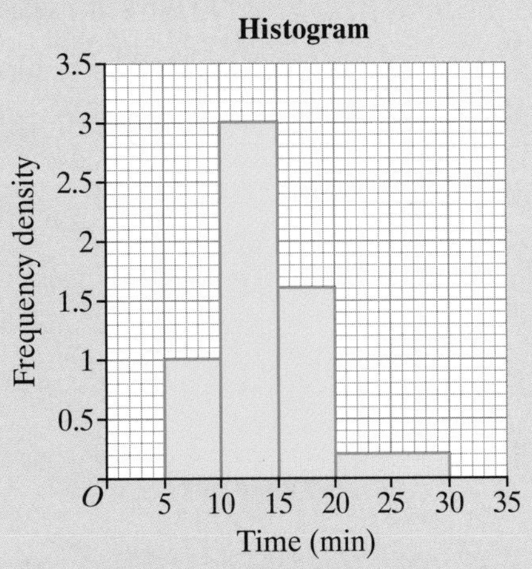

Frequency polygon – this is drawn from a histogram (or bar chart) by joining the midpoints of the tops of the bars with straight lines to form a polygon

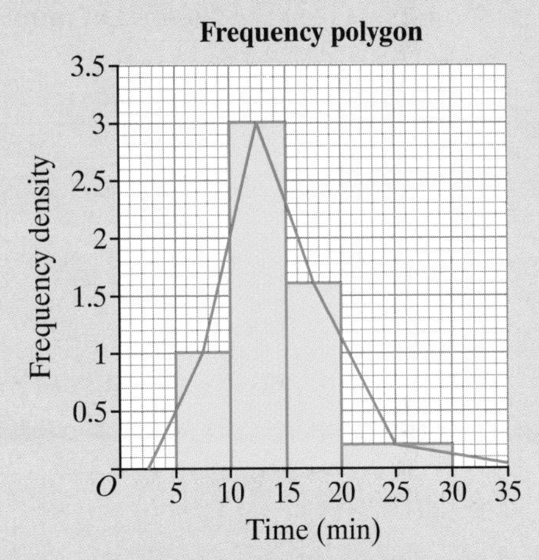

Frequency density – in a histogram, the area of the bars represents the frequency and the height represents the frequency density

$$\text{Frequency density (bar height)} = \frac{\text{frequency}}{\text{class width}}$$

Learn 1 Stem-and-leaf diagrams

Examples:

a The number of minutes taken to complete an exercise was recorded for 15 boys in a class.

16, 27, 28, 22, 34, 18, 11, 19, 27, 31, 27, 36, 22, 17, 29

Show the information as a stem-and-leaf diagram.

Number of minutes to complete a task

Stem (tens)	Leaf (units)
1	6 8 1 9 7
2	7 8 2 7 7 2 9
3	4 1 6

In this case the number 6 stands for 16 (1 ten and 6 units)

In this case the number 6 stands for 36 (3 tens and 6 units)

Key: 3|4 represents 34 minutes

It is useful to provide an ordered stem-and-leaf diagram to find the median and range.

Number of minutes to complete a task

Stem (tens)	Leaf (units)
1	1 6 7 8 9
2	2 2 7 7 7 8 9
3	1 4 6

Here the leaves (units) are all arranged numerically

Key: 3|4 represents 34 minutes

b The number of minutes taken to complete an exercise was recorded for 15 boys and 15 girls in a class.

Boys: 16, 27, 28, 22, 34, 18, 11, 19, 27, 31, 27, 36, 22, 17, 29
Girls: 12, 23, 22, 17, 30, 16, 15, 14, 17, 37, 26, 24, 21, 12, 27

Show the information as a back-to-back stem-and-leaf diagram.

Here the leaves (units) are all arranged numerically from the right-hand side

Number of minutes to complete a task

Leaf (units) Girls	Stem (tens)	Leaf (units) Boys
7 7 6 5 4 2 2	1	1 6 7 8 9
7 6 4 3 2 1	2	2 2 7 7 7 8 9
7 0	3	1 4 6

Key: 3|2 represents 23 minutes

Key: 3|2 represents 32 minutes

The boys' data has already been recorded in an ordered stem-and-leaf diagram

This diagram is called a back-to-back (ordered) stem-and-leaf diagram.

Apply 1

1 The prices paid for some takeaway food are shown below.

£3.64 £4.15 £5.22 £5.88 £4.21 £4.55 £3.75 £4.78 £5.05 £4.52 £4.60

Copy and complete the stem-and-leaf diagram to show this information.

Prices paid for takeaway food

Stem (£)	Leaf (pence)
3	
4	
5	

Key: 4 | 15 represents £4.15

2 The marks obtained in a test were recorded as follows.

8 20 9 21 18 22 19 13 22 24 14 9 25 10 19 20 17 14 12

a Show this information in an ordered stem-and-leaf diagram.

b What was the highest mark in the test?

c Write down the median of the marks in the test.

d Write down the range of the marks in the test.

3 The times taken to complete an exam paper were:

2 h 12 min, 1 h 53 min, 1 h 26 min, 2 h 26 min, 1 h 50 min, 1 h 46 min, 2 h 05 min, 1 h 43 min, 1 h 49 min, 2 h 10 min, 1 h 49 min, 1 h 55 min, 2 h 06 min, 1 h 57 min.

Show this information in an ordered stem-and-leaf diagram.

4 The heights of some students are shown in this stem-and-leaf diagram.

Heights of students

Stem (feet)	Leaf (inches)
4	10 09
5	08 07 03 11 08 01 10 00 08 07 06
6	02 01

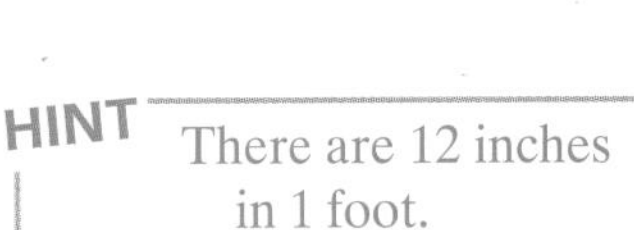

HINT There are 12 inches in 1 foot.

Key: 5 | 06 represents 5 feet 6 inches

a Rearrange the diagram to produce an ordered stem-and-leaf diagram.

b Use your diagram to answer these questions.

i How many students were included altogether?

ii Find the mode.

iii Find the median.

iv Calculate the range.

5 The number of minutes taken to complete an exercise was recorded for 15 boys and 15 girls in this back-to-back stem-and-leaf diagram.

Number of minutes to complete a task

Leaf (units) Girls	Stem (tens)	Leaf (units) Boys
7 7 6 5 4 2 2	1	1 6 7 8 9
7 6 4 3 2 1	2	2 2 7 7 7 7 8 9
7 0	3	1 4 6

Key : 3|2 represents 23 minutes   Key : 3|4 represents 34 minutes

a Calculate the median for the girls.

b Calculate the mode for the boys.

c Calculate the mean for the girls.

d Calculate the mean for the boys.

e Calculate the range for the girls.

f Calculate the range for the boys.

g What can you say about the length of time to complete the exercise by the girls and the boys?

6 Jenny records the reaction times of students in Year 7 and Year 11 at her school.

Year	Times (tenths of a second)														
7	18	19	09	28	10	04	11	14	15	18	09	27	28	06	05
11	07	20	09	12	21	17	11	12	15	08	09	12	08	16	19

a Show this information in a back-to-back stem-and-leaf diagram.

b Jenny's hypothesis is 'The reaction times of Year 7 students are quicker than that of Year 11 students.'
Use your data to check Jenny's hypothesis.

Explore

Collect some information from the students in your class, and show it in a stem-and-leaf diagram. You might wish to collect:

- Height, arm length, head circumference ... in metric or imperial units
- Estimate of the weight of ten mathematics books in metric or imperial units
- Time to complete a task in minutes and seconds
- Time spent on homework in hours and minutes
- Time to run or walk a certain distance in minutes and seconds
- Weekly pocket money or weekly wages in pounds and pence

Alternatively, collect some other information that can be shown in a stem-and-leaf diagram

Investigate further

Learn 2 Frequency diagrams, line graphs and time series

Examples:

a 50 people were asked how long they had to wait for a train. The table below shows the results.

Time, t (minutes)	Frequency
$5 \leqslant t < 10$	16
$10 \leqslant t < 15$	22
$15 \leqslant t < 20$	11
$20 \leqslant t < 25$	1

Draw a frequency diagram to represent the data.

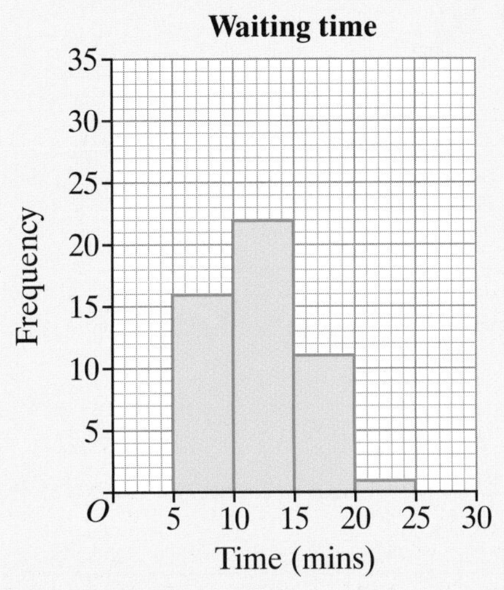

A frequency diagram is similar to a bar chart except that it is used for continuous data

b The table shows the temperature of a patient at different times during the day.

Time	10.00	11.00	12.00	13.00	14.00
Temperature (°F)	102.5	101.3	102	99.1	99.2

Draw a line graph to show this information.

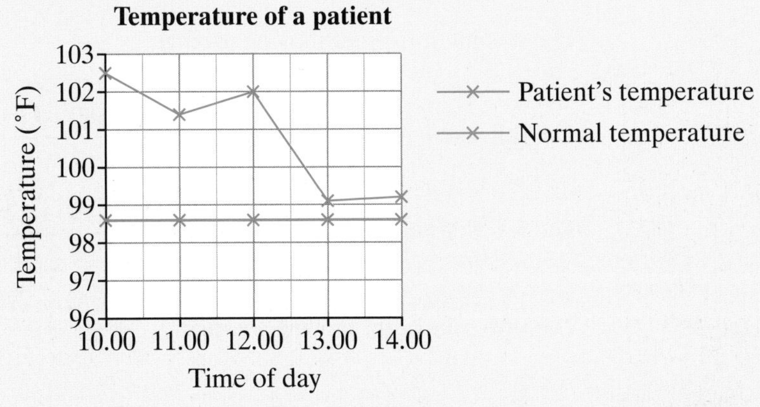

In this example, the temperature is expected to fall to the normal body temperature of 98.8°F. However, other line graphs can fluctuate over time.

c The table shows the cost of gas bills at the end of every three months.

Year	2006	2006	2006	2006	2007	2007	2007	2007
Quarter	1st	2nd	3rd	4th	1st	2nd	3rd	4th
Cost	£260	£170	£80	£100	£280	£190	£110	£140

i Draw a time series to show this information.

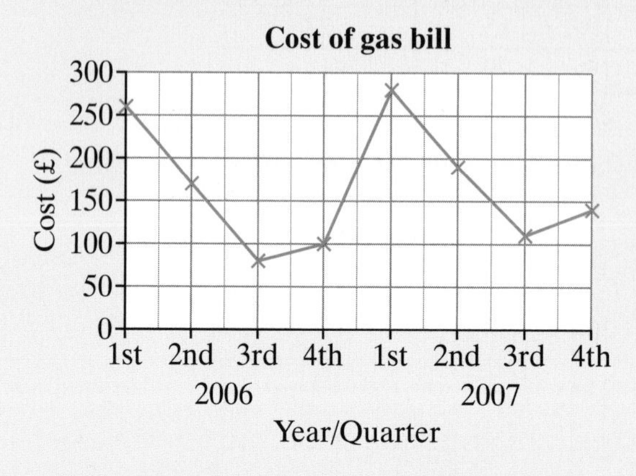

ii Use the table to find the moving average and plot this on your graph. Comment on the trend of your graph.

In this example, the cost follows a cycle that seems to be repeating every four quarters, so you can use a four-point moving average to find any trend

To find the four-point moving average take the mean of each four successive points.

Year	2006	2006	2006	2006	2007	2007	2007	2007
Quarter	1st	2nd	3rd	4th	1st	2nd	3rd	4th
Cost	£260	£170	£80	£100	£280	£190	£110	£140

The first four-point moving average $= \dfrac{£260 + £170 + £80 + £100}{4} = £152.50$

The second four-point moving average $= \dfrac{£170 + £80 + £100 + £280}{4} = £157.50$

The third four-point moving average $= \dfrac{£80 + £100 + £280 + £190}{4} = £162.50$

The fourth four-point moving average $= \dfrac{£100 + £280 + £190 + £110}{4} = £170.00$

The fifth four-point moving average $= \dfrac{£280 + £190 + £110 + £140}{4} = £180.00$

To find the second moving average, miss out the first reading and include the fifth reading

Plotting the four-point moving averages on the graph:

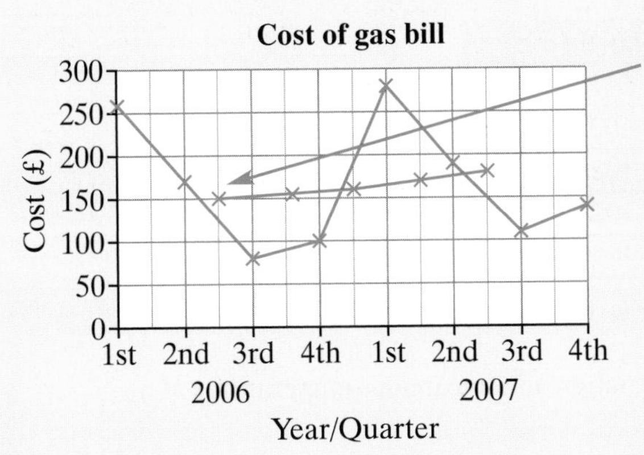

The first four-point moving average is plotted in the 'middle' of the first four points, and so on

From this graph you can see that the trend is upwards.

Apply 2

1 The table shows the time spent in a local shop by 60 customers.

Time, t (minutes)	Frequency
$5 \leqslant t < 10$	8
$10 \leqslant t < 15$	30
$15 \leqslant t < 20$	16
$20 \leqslant t < 25$	6

Draw a frequency diagram to represent the data.

2 The frequency diagram shows the ages of 80 people in a factory.

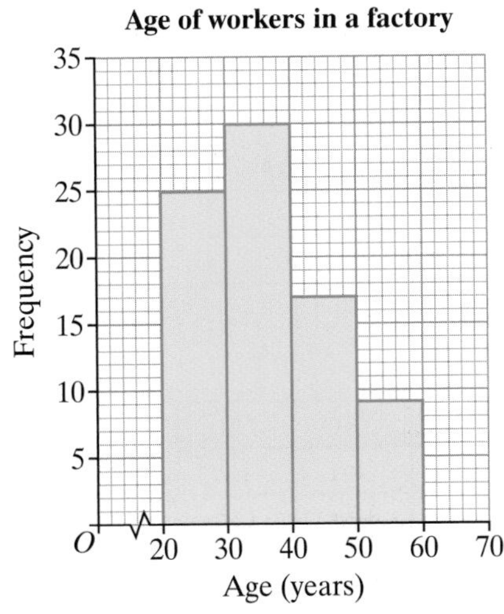

Copy and complete this table to show this information.

Age, y (years)	Frequency
$20 \leqslant y < 30$	
$30 \leqslant y < 40$	
$40 \leqslant y < 50$	
$50 \leqslant y < 60$	

3 The table shows the pressure in millibars (mb) over five days at a seaside resort.

Day	Pressure (mb)
Monday	1018
Tuesday	1022
Wednesday	1028
Thursday	1023
Friday	1019

Draw a line graph to show the pressure each day.

4 The table shows the minimum and maximum temperatures at a seaside resort.

Day	Minimum temperature (°C)	Maximum temperature (°C)
Monday	15	19
Tuesday	11	21
Wednesday	13	22
Thursday	14	23
Friday	17	23

Draw a line graph to show:

a the minimum temperatures

b the maximum temperatures.

Use your graph to find:

c the day on which the lowest temperature was recorded

d the day on which the highest temperature was recorded

e the biggest difference between the daily minimum and maximum temperatures.

5 The table shows the cost of electricity bills at the end of every three months.

Year	2006	2006	2006	2006	2007	2007	2007	2007
Quarter	1st	2nd	3rd	4th	1st	2nd	3rd	4th
Cost	£230	£120	£50	£80	£215	£120	£25	£55

a Copy this table and calculate the four-point moving averages. The first two have been done for you.

Year	2006	2006	2006	2006	2007	2007	2007	2007
Quarter	1st	2nd	3rd	4th	1st	2nd	3rd	4th
Cost	£230	£120	£50	£80	£215	£120	£25	£55
Four-point moving average		£120.00	£116.25					

b Show this information on a graph.

c What can you say about the trend?

6 The table shows the number of students at college present during morning and afternoon registration.

Day	Mon	Mon	Tue	Tue	Wed	Wed	Thu	Thu	Fri	Fri
Session	a.m.	p.m.	a.m.	p.m.	a.m.	p.m.	a.m.	p.m.	a.m.	p.m.
Number	220	210	243	215	254	218	251	201	185	152

a Copy this table and calculate the two-point moving averages. The first four have been done for you.

Day	Mon	Mon	Tue	Tue	Wed	Wed	Thu	Thu	Fri	Fri
Session	a.m.	p.m.	a.m.	p.m.	a.m.	p.m.	a.m.	p.m.	a.m.	p.m.
Number	220	210	243	215	254	218	251	201	185	152
Two-point moving average	215	226.5	229	234.5						

b Show this information on a graph.

c What can you say about the trend?

7 Simon keeps a note of his termly exam results.

Year	2005	2006	2006	2006	2007	2007	2007
Session	Autumn	Spring	Summer	Autumn	Spring	Summer	Autumn
Exam result (%)	86	93	70	83	93	67	77

a Use the information to calculate the three-point moving averages.

b Show this information on a graph.

c Simon says that his performance is improving. Is he correct? Give a reason for your answer.

8 The graph shows the quarterly sales at a shop.

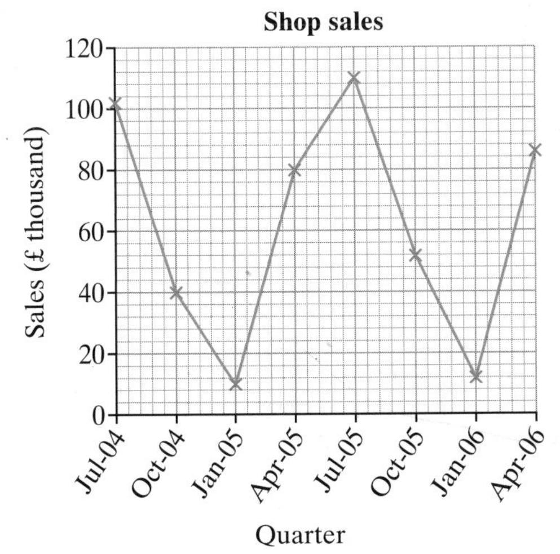

a Use the graph to calculate the four-point moving averages.

b Copy the graph and plot your four-point moving averages.

c What can you say about the sales trend?

d Use the moving averages to predict the likely sales for July 2006.

9 The table shows a company's quarterly profits (£ million) recorded at the end of each quarter in 2005 and 2006.

Year	Quarter	Profit (£ million)	Four-point moving average
2005	March	1.2	
	June	1.5	1.35
	September	1.9	
	December	0.8	
2006	March	1.1	
	June	1.3	
	September	2.0	
	December	1.1	

a Calculate the four-point moving averages.
The first one has been done for you.

b Show this information on a graph.

c Does the graph suggest that profits are going up or down?
Give a reason for your answer.

d Use the trend line to predict the likely profit for March 2007.

10 Copy and complete the table. Identify the most appropriate moving average for each item. One of them has been done for you.

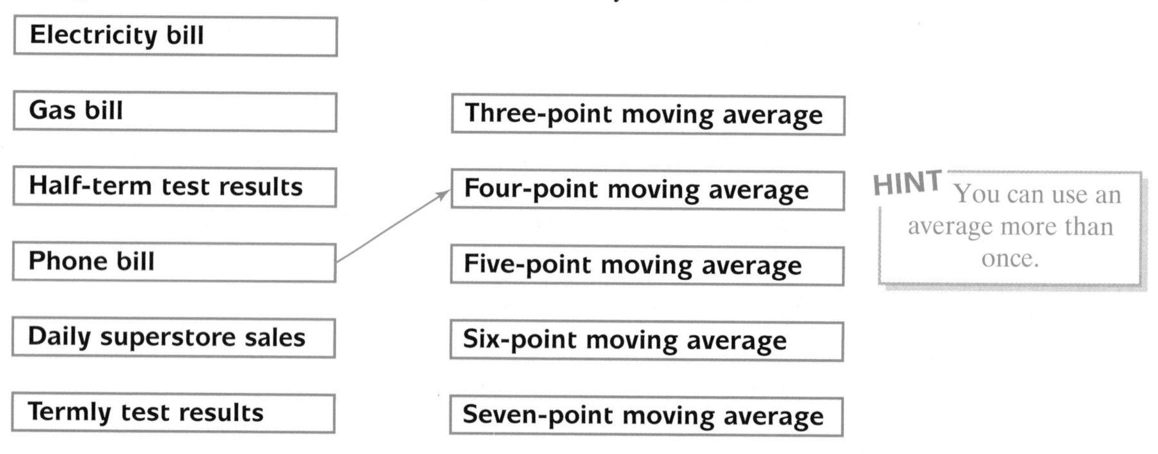

HINT You can use an average more than once.

Explore

Collect old bills together and see how they change over time. Use this information to create your own time series. Work out the moving averages – which is the most appropriate moving average?

Suggestions include:

- Gas bills
- Electricity bills
- Water bills
- Telephone/mobile bills
- Pay slips

Investigate further

Explore

Collect information over a period of two weeks that you can use to create your own time series. Work out the moving averages – which is the most appropriate moving average?

Suggestions for data include:

- Time taken to complete homework
- Time taken to travel to work/school/college
- Time spent sleeping/eating/exercising
- Homework/examination or test results over time

Investigate further

Learn 3 Cumulative frequency diagrams

Example:

The table shows the times taken to complete 20 telephone calls.

Time (min)	Frequency
$0 \leqslant t < 3$	5
$3 \leqslant t < 6$	8
$6 \leqslant t < 9$	4
$9 \leqslant t < 15$	3

The group $3 \leqslant t < 6$ means greater than or equal to 3 and less than 6

The value of 6 minutes will not be included in this interval

Show this information on a cumulative frequency diagram.
Use your diagram to estimate the median and interquartile range.

To find cumulative frequencies, you add the frequencies in turn to give you a 'running total'.

Add a third column for the cumulative frequency.

Time (min)	Frequency	Cumulative frequency
$0 \leqslant t < 3$	5	5
$3 \leqslant t < 6$	8	13
$6 \leqslant t < 9$	4	17
$9 \leqslant t < 15$	3	20

5 calls took less than 3 minutes

$5 + 8 = 13$ calls took less than 6 minutes

Check that the final cumulative frequency is the same as the total number of calls

Now you can draw the cumulative frequency diagram.

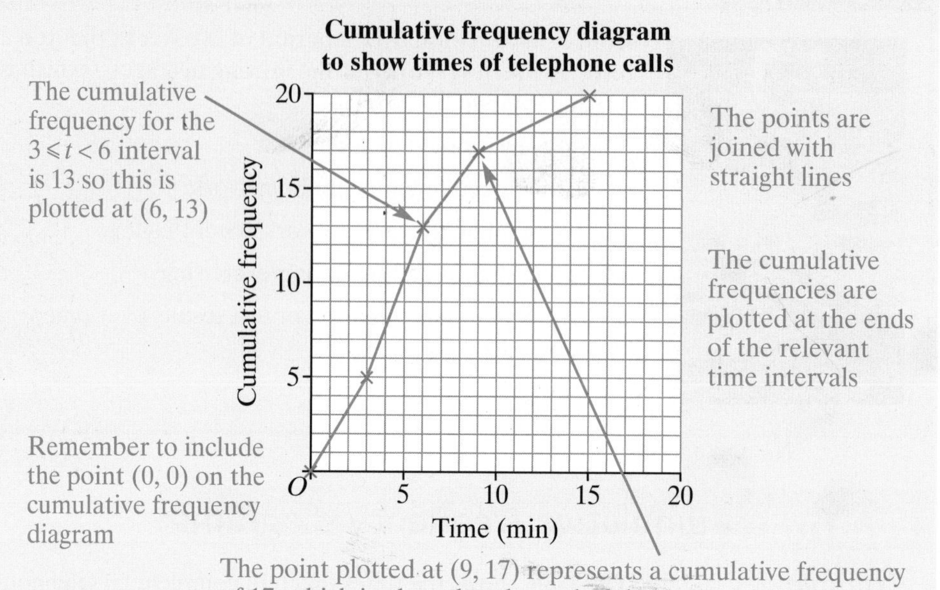

The cumulative frequency can be divided by four to find the quartiles and the median as follows.

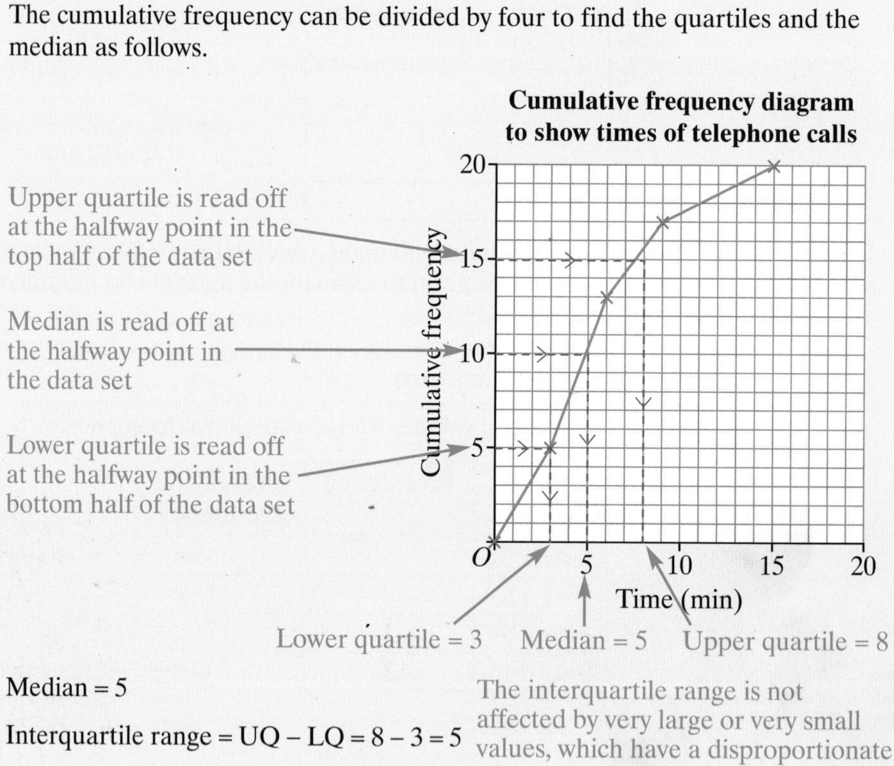

Median = 5

Interquartile range = UQ – LQ = 8 – 3 = 5

The interquartile range is not affected by very large or very small values, which have a disproportionate effect on the range

Apply 3

1 The table shows the distances travelled to work by 40 commuters.

Distance (miles)	Frequency
$0 < d \leqslant 2$	11
$2 < d \leqslant 4$	16
$4 < d \leqslant 8$	10
$8 < d \leqslant 16$	3

a Draw a cumulative frequency diagram to show this data.

b Use your diagram to estimate:

i the median distance

ii the interquartile range.

2 The table shows the heights of 50 students in a class.

Height (cm)	Frequency
$150 < h \leqslant 155$	4
$155 < h \leqslant 160$	7
$160 < h \leqslant 165$	18
$165 < h \leqslant 170$	11
$170 < h \leqslant 175$	6
$175 < h \leqslant 180$	4

a Show this information on a cumulative frequency diagram.

b Use your diagram to estimate:

i the median height

ii the limits between which the middle 50% of the heights lie.

3 The table shows the wages of workers in a factory.

Wages (£)	Frequency
$0 < x \leqslant 100$	120
$100 < x \leqslant 150$	165
$150 < x \leqslant 200$	182
$200 < x \leqslant 250$	197
$250 < x \leqslant 300$	40
$300 < x \leqslant 500$	6

a Draw a cumulative frequency diagram to show this data.

b Use your diagram to estimate:

i the median wage

ii the interquartile range

iii the number of workers whose wage is below £175

iv the number of workers whose wage is above £420.

HINT First find the number who earn £420 or less and then subtract from the total number of workers.

4 Peter records the waiting times (to the nearest minute) at a post office and puts the times in a table.

Time (min)	Frequency
1–3	23
4–6	17
7–9	8
10–15	2

a Show this information on a cumulative frequency diagram.

b Use your diagram to estimate:

i the median waiting time

ii the interquartile range

iii the percentage of people who waited over twelve minutes.

5 The cumulative frequency diagram shows the times taken by 60 students to complete an arithmetic test.

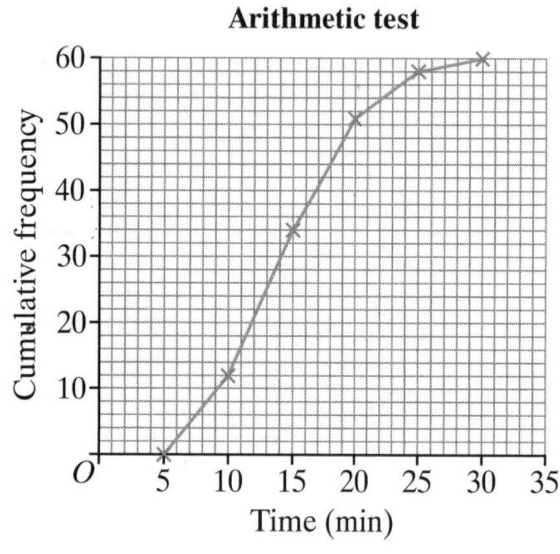

a Copy and complete the table to show the cumulative frequencies.

Time (min)	Cumulative frequency
$5 < t \leqslant 10$	
$10 < t \leqslant 15$	
$15 < t \leqslant 20$	
$20 < t \leqslant 25$	
$25 < t \leqslant 30$	

b Use your table to complete the frequency distribution.

Time (min)	Frequency
$5 < t \leqslant 10$	
$10 < t \leqslant 15$	
$15 < t \leqslant 20$	
$20 < t \leqslant 25$	
$25 < t \leqslant 30$	

6 The table shows the arm spans of a class of 25 students.

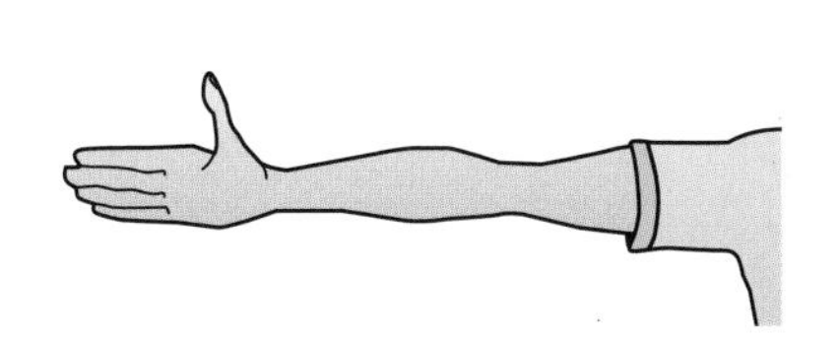

Arm span (cm)	Frequency
160–170	0
170–180	3
180–190	11
190–200	9
200–210	2

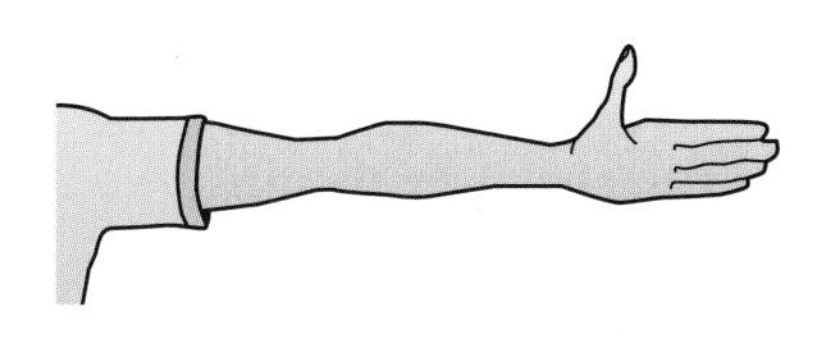

a Show this information on a cumulative frequency diagram.

b Use your diagram to estimate:

i the median

ii the interquartile range.

Another class of 25 students has this distribution.

Median	208
Interquartile range	26

c What can you say about the two classes?
Give reasons for your answer.

7 The table shows the number of words per paragraph in a newspaper.

Words	Number of paragraphs
0	0
1–10	15
11–20	31
21–30	45
31–40	25
41–50	4

a Show this information on a cumulative frequency diagram.

b Use your diagram to estimate:

i the median number of words

ii the interquartile range.

This table shows the number of words per paragraph in a magazine.

Words	Number of paragraphs
0	0
1–10	20
11–20	46
21–30	36
31–40	13
41–50	5

c Draw a cumulative frequency diagram on the same graph as part **a**.
What can you say about the two distributions?
Give reasons for your answer.

8 Match the frequency diagram with the cumulative frequency diagram.

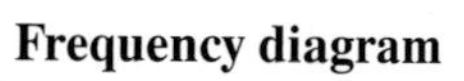

Frequency diagram

a

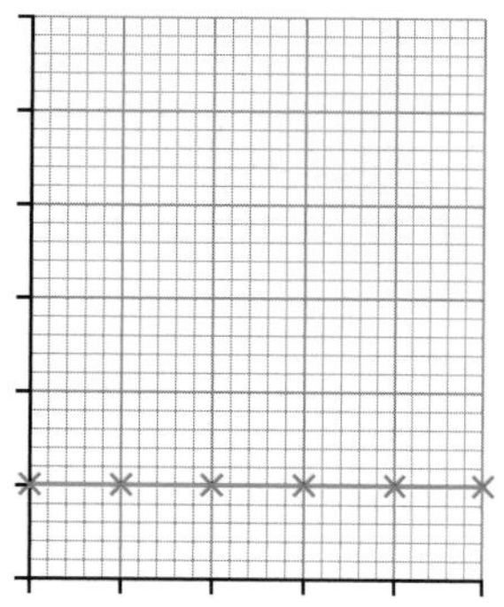

b

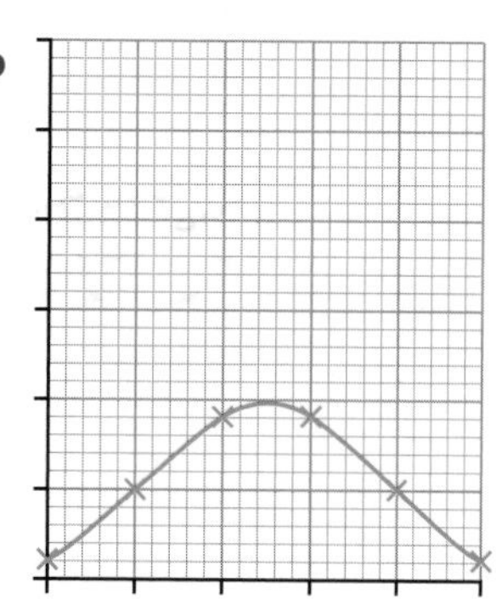

c

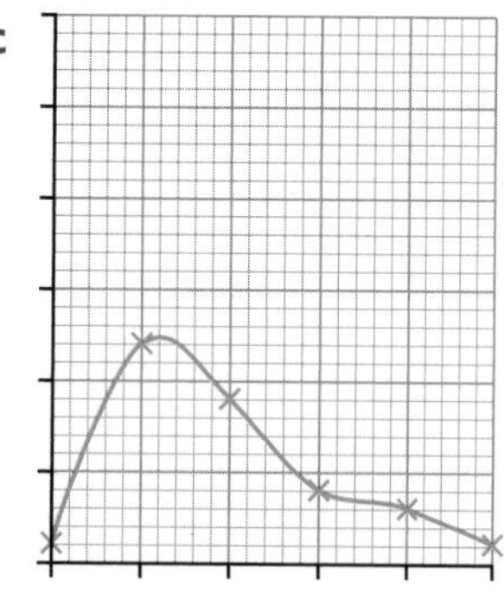

d

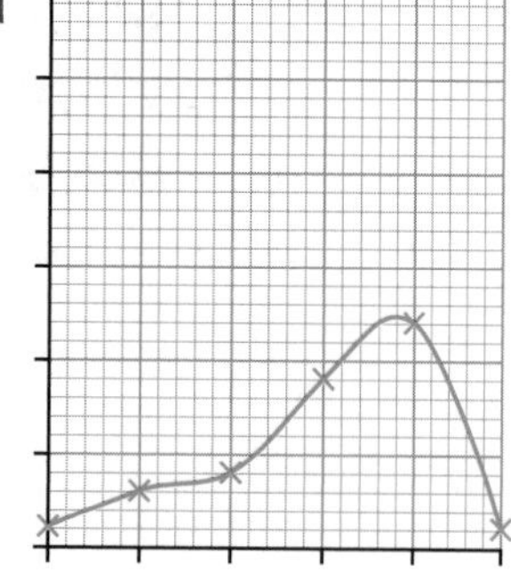

Cumulative frequency diagram

i

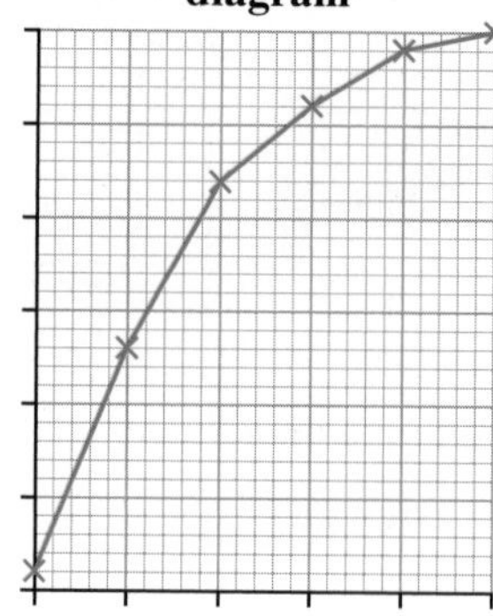

ii

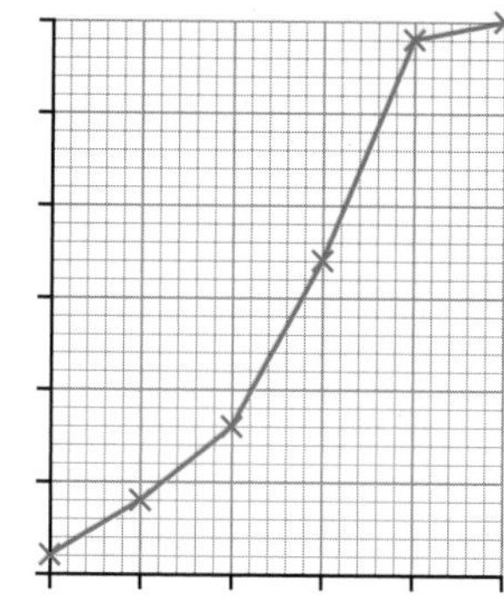

iii

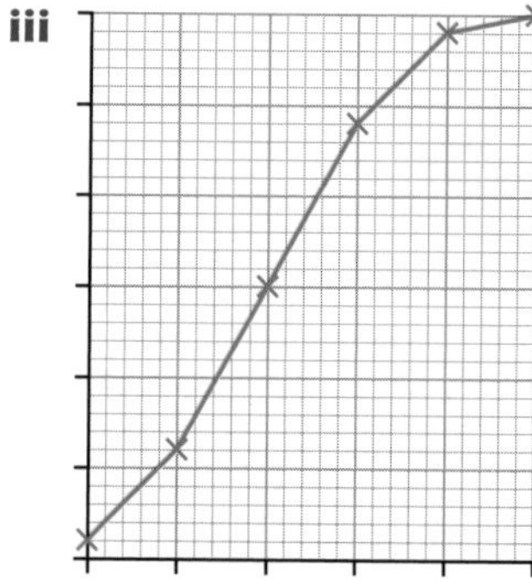

iv

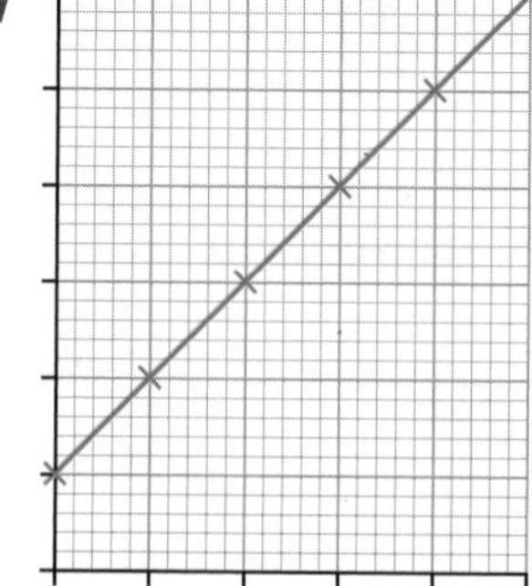

Explore

Collect data from your work/school/family/internet about these variables:

- Height
- Arm length
- Head circumference
- Running time over a fixed distance
- Pulse rate
- Heart rate
- Time taken to complete homework
- Time taken to travel to work/school/college
- Time spent sleeping/eating/exercising

Use your data to draw cumulative frequency diagrams
Where do you lie in each distribution?

Investigate further

Learn 4 Box plots

Examples:

a Peter keeps a note of his sales over the past nine days.

Sun	Mon	Tue	Wed	Thu	Fri	Sat	Sun	Mon
28	2	8	11	16	15	30	25	4

Draw a box plot for the data.

To draw a box plot you need:

- the minimum and maximum values
- the lower and upper quartiles
- the median.

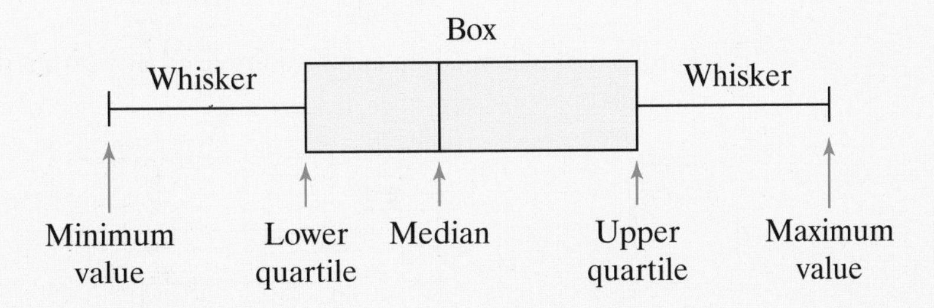

Rearranging the data in order:

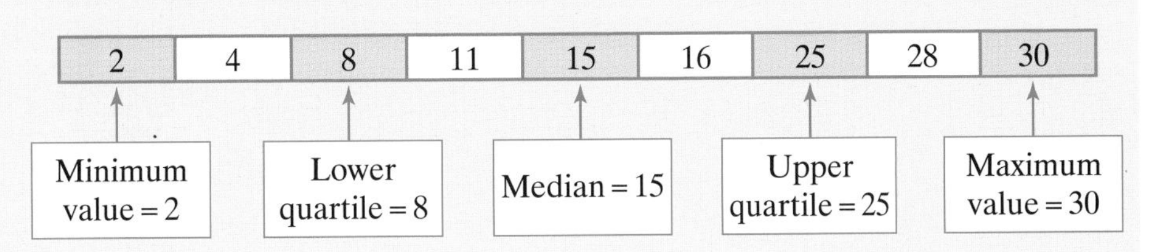

The box plot looks like this:

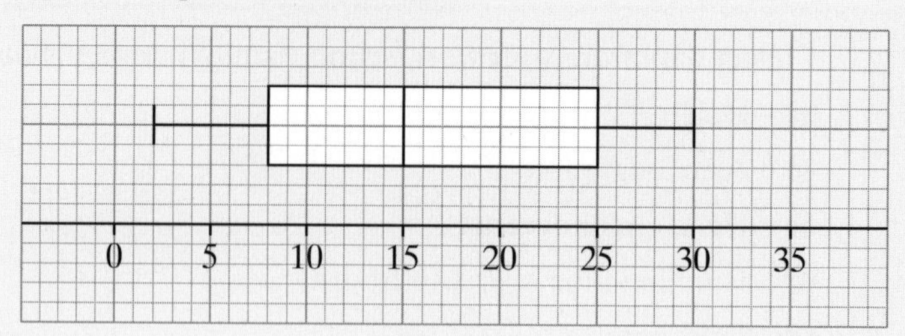

Sales (£)

b Mary also keeps a note of her sales over the past nine days.

Sun	Mon	Tue	Wed	Thu	Fri	Sat	Sun	Mon
26	3	10	18	14	21	19	31	6

Draw a box plot for the data.

Rearranging the data in order:

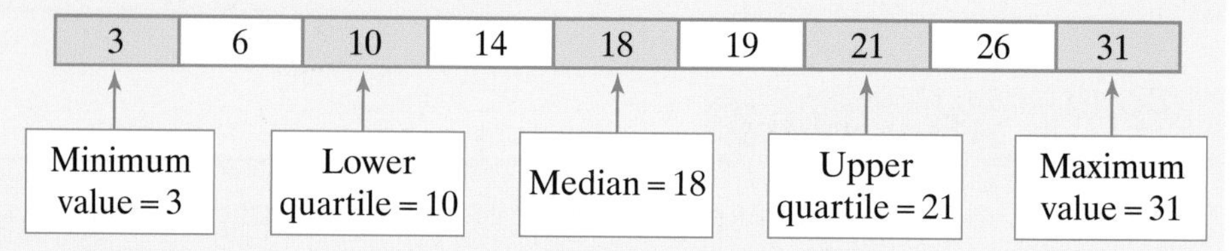

Mary's box plot:

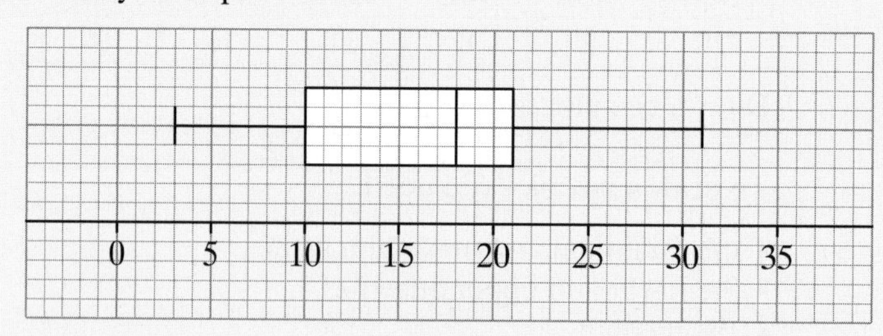

Sales (£)

Peter's and Mary's data can be compared by putting the two box plots on the same axis.

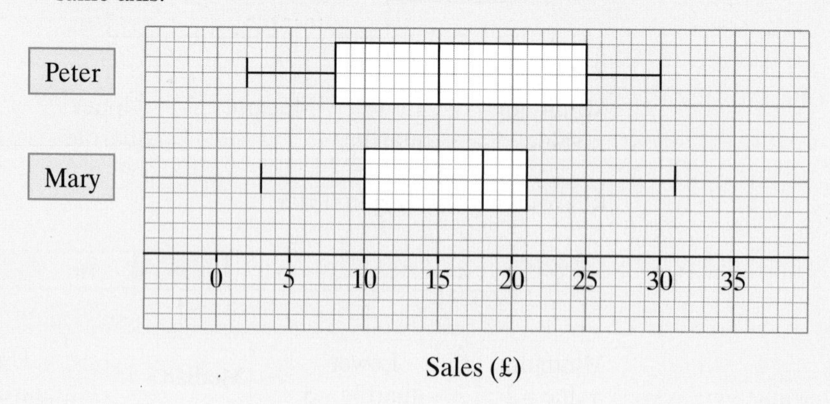

Sales (£)

Information from graphs should always be interpreted in terms of the given data

Measure	Description	Interpretation
Median	Mary's median is greater	Mary's average (median) sales are greater than Peter's average (median) sales
IQR	Peter's interquartile range is greater	Taking the middle 50% of the sales, Peter's sales are more spread out than Mary's sales
Range	The ranges are the same	Over the whole range, the spread of sales is the same (£28) for both Mary and Peter

Box plots can also be used for highlighting and comparing information from cumulative frequency diagrams.

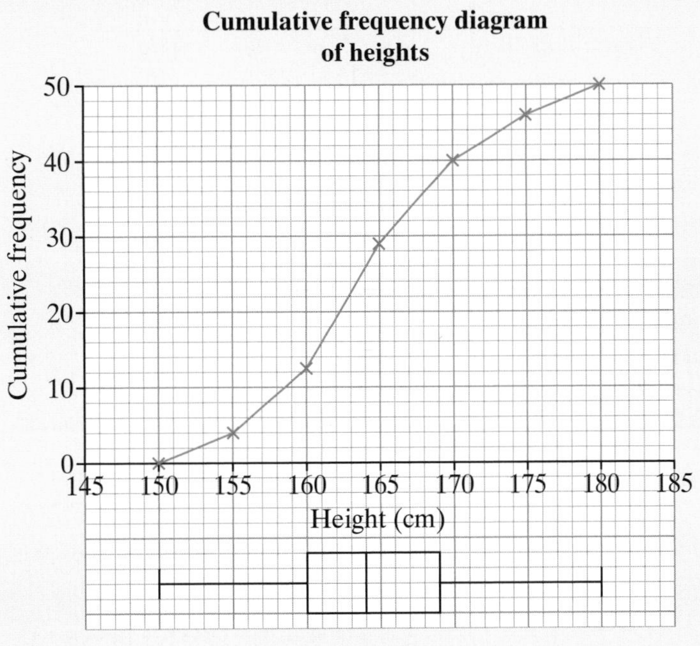

From the cumulative frequency diagram:

- the minimum value = 150
- the maximum value = 180
- the lower quartile = 160
- the upper quartile = 169
- the median = 164

Apply 4

1 Draw box plots of this data.

a

23	7	7	11	28	17	9	21	8	18	25	29	5	13	25

b

13	5	13	17	20	15	3	7	11	8	9

c

12	5	14	15	20	24	4	6	11	6	9	15

2 These box plots show the ages of all of the people in two villages.

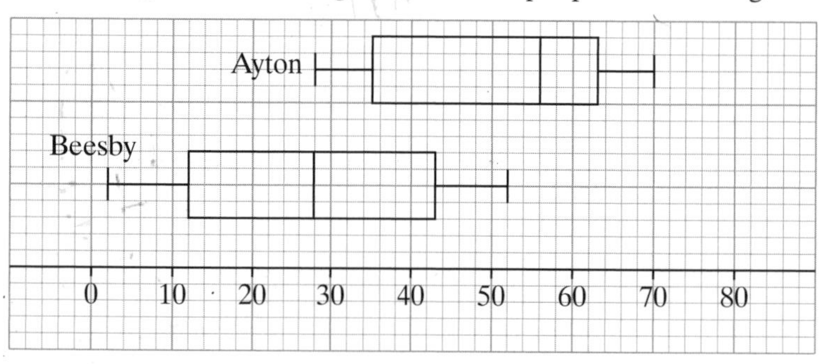

Age (years)

a Copy and complete the table for the two villages.

	Ayton	Beesby
Minimum age		
Maximum age		
Lower quartile		
Upper quartile		
Median		

b Compare and contrast the ages of people in the two villages.

3 The examination results for 60 students are shown in the table.

1–10	11–20	21–30	31–40	41–50	51–60	61–70	71–80	81–90	91–100
1	1	2	3	5	9	10	14	12	3

a Show this information on a cumulative frequency diagram.

b Use the diagram to estimate:

i the median

ii the lower quartile

iii the upper quartile.

c Draw a box plot for the data.

4 Match each cumulative frequency diagram with a box plot.

a

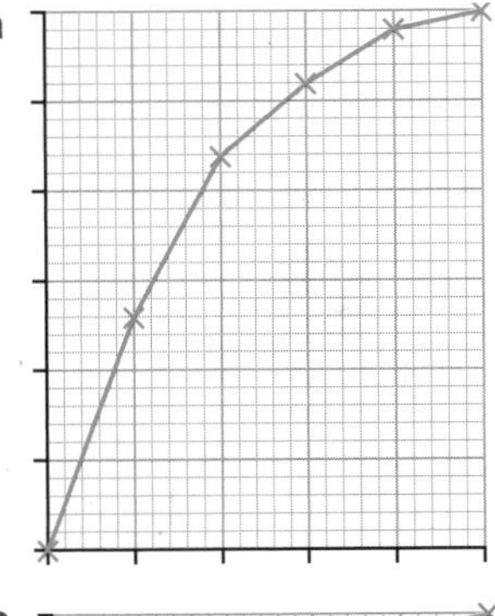

i

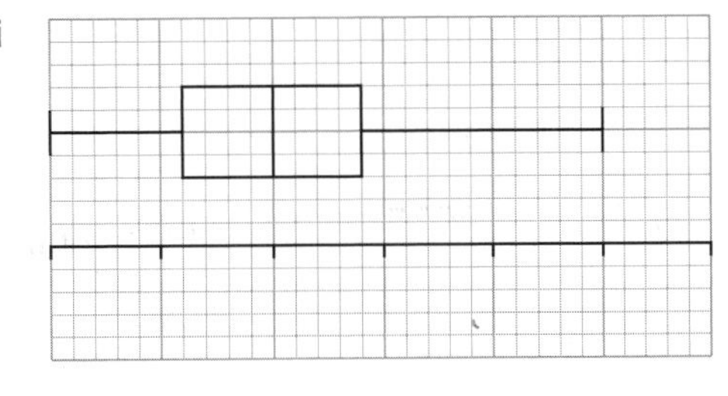

b

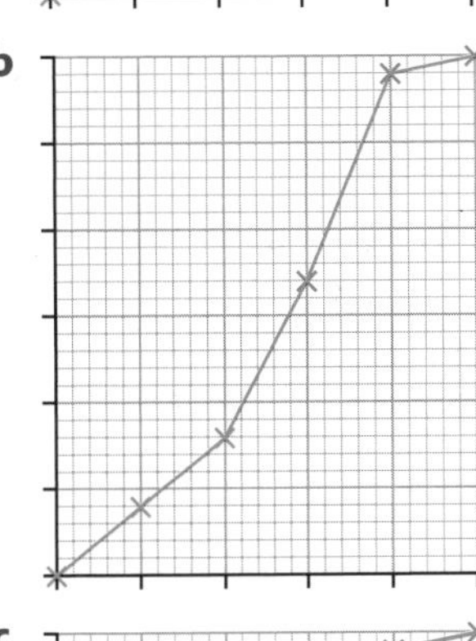

ii

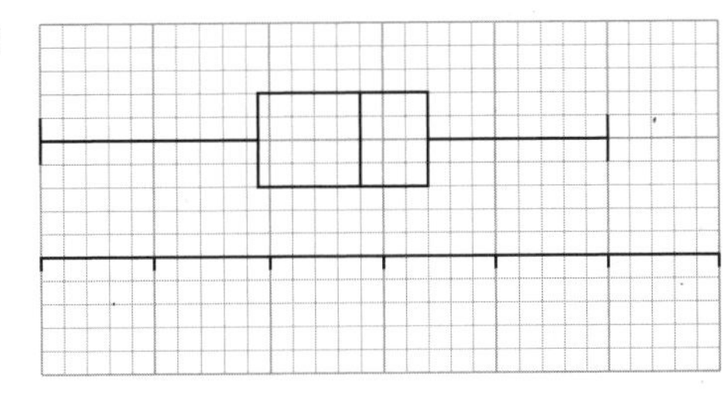

c

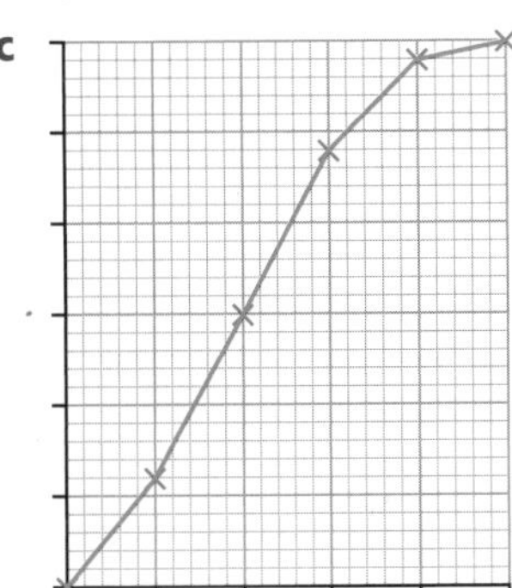

iii

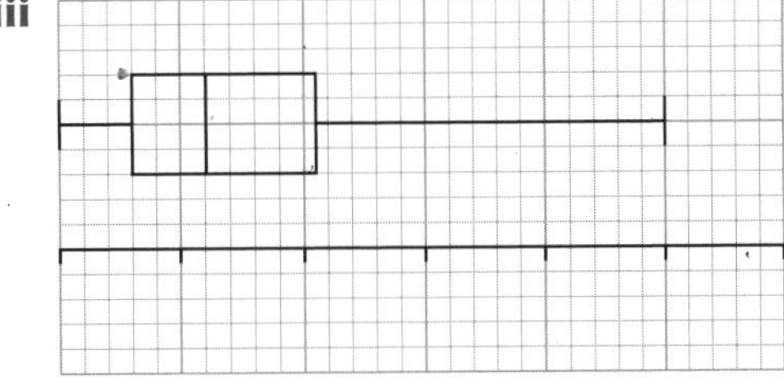

5 The cumulative frequency diagram shows the waiting times (to the nearest minute) at a main post office.

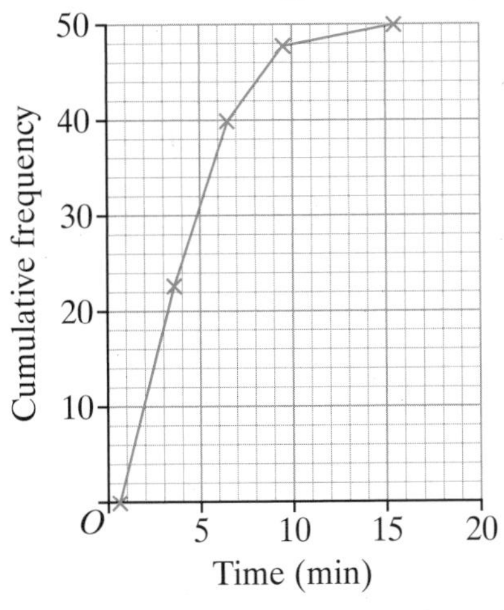

a Use the diagram to draw a box plot for the data.

b This box plot shows the waiting times at a village post office. Write down two differences between the waiting times at the two post offices.

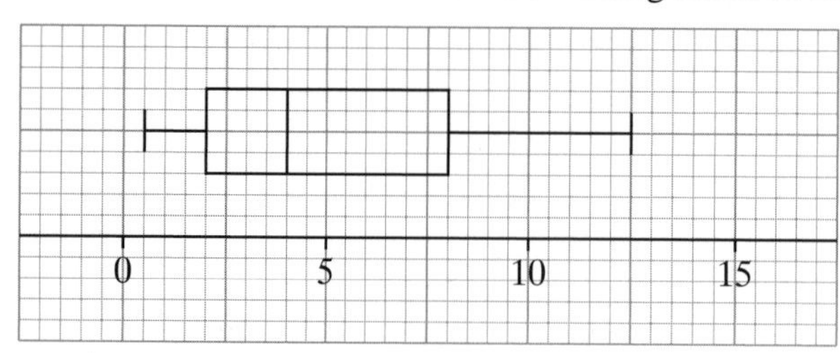

Waiting time (min)

Explore

Collect data from your work/school/family/internet about these variables:

- Height
- Arm length
- Head circumference
- Running time over a fixed distance
- Pulse rate
- Heart rate
- Time taken to complete homework
- Time taken to travel to work/school/college
- Time spent sleeping/eating/exercising

Use your data to draw box plots. Where do you lie in each distribution?

Investigate further

Learn 5 Histograms

Examples:

The table shows the waiting times in an accident and emergency department.

Time (min)	Frequency
$5 \leqslant t < 10$	5
$10 \leqslant t < 15$	15
$15 \leqslant t < 20$	8
$20 \leqslant t < 30$	2

In a histogram, the area of the bar represents the frequency

Frequency = class width × height of bar

or

$$\text{Height} = \frac{\text{frequency}}{\text{class width}}$$

Show the information on:

a a histogram
b a frequency polygon.

a Add additional columns for class width and height (frequency density).

Time (min)	Frequency	Class width	Height = $\frac{\text{frequency}}{\text{class width}}$
$0 \leqslant t < 5$	0	5	Height = $0 \div 5 = 0$
$5 \leqslant t < 10$	5	5	Height = $5 \div 5 = 1$
$10 \leqslant t < 15$	15	5	Height = $15 \div 5 = 3$
$15 \leqslant t < 20$	8	5	Height = $8 \div 5 = 1.6$
$20 \leqslant t < 30$	2	10	Height = $2 \div 10 = 0.2$

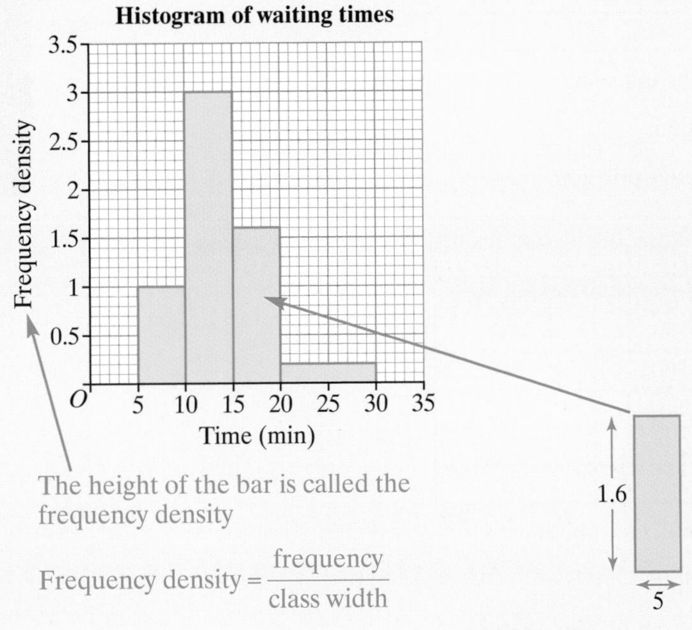

Width = 5
Height = 1.6
Frequency = class width × height
= 5 × 1.6
= 8

The bar represents a frequency of 8.

b A frequency polygon can be drawn from a histogram (or bar chart) by joining the midpoints of the tops of the bars with straight lines to form a polygon.

The area of the polygon is equal to the total frequency.

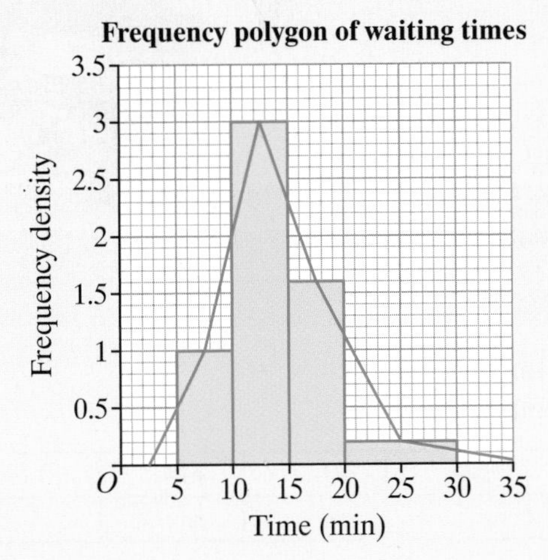

It is usual to extend the lines to the horizontal axis as shown

Apply 5

1 The table shows the distances travelled to work by 36 commuters.

Distance (miles)	Frequency
$0 < d \leqslant 1$	0
$1 < d \leqslant 2$	6
$2 < d \leqslant 4$	16
$4 < d \leqslant 8$	11
$8 < d \leqslant 16$	3

Illustrate the data on:

a a histogram

b a frequency polygon.

2 The table shows the arm spans for 25 students in a class.

Arm span (cm)	Frequency
160–180	0
180–200	3
200–210	11
210–220	9
220–240	2

Show this information on:

a a histogram

b a frequency polygon.

3 The ages of people at a hotel are shown in the histogram below.

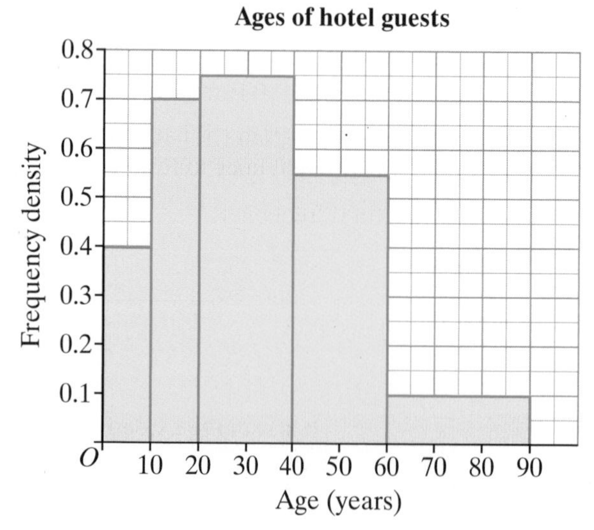

Copy and complete the table for the distribution.
The first one has been done for you.

Age (years)	0–	10–	20–	40–	60–90
Number of guests	4				

4 The weights of 60 students in a year group were recorded.

Weight (kg)	Frequency
$40 \leqslant w < 42$	1
$42 \leqslant w < 45$	9
$45 \leqslant w < 50$	22
$50 \leqslant w < 55$	18
$55 \leqslant w < 65$	8
$65 \leqslant w < 75$	2

Show this information on:

a a histogram

b a frequency polygon.

5 Light bulbs are tested to see how long they last.
The table shows the results of 60 tests.

Time (hours)	Frequency
$600 \leqslant x < 700$	3
$700 \leqslant x < 800$	15
$800 \leqslant x < 850$	20
$850 \leqslant x < 900$	12
$900 \leqslant x < 1000$	8
$1000 \leqslant x < 1200$	2

a Draw a histogram to show this information.

b Use your histogram to estimate the median life of a light bulb.

HINT Draw a vertical line dividing the area of the histogram into two equal parts.

6 The table shows the wages of workers in a factory.

Wages (£)	Frequency
$0 \leqslant x < 100$	120
$100 \leqslant x < 150$	165
$150 \leqslant x < 200$	182
$200 \leqslant x < 250$	197
$250 \leqslant x < 300$	40
$300 \leqslant x < 500$	6

Show this information on:

a a histogram

b a frequency polygon.

Explore

Collect data from your work/school/family about these variables:

- Height
- Arm length
- Head circumference
- Running time over a fixed distance
- Pulse rate
- Heart rate
- Time taken to complete homework
- Time taken to travel to work/school/college
- Time spent sleeping/eating/exercising

Use your data to draw histograms. Where do you lie in each distribution?

Investigate further

Representing data

ASSESS

The following exercise tests your understanding of this chapter, with the questions appearing in order of increasing difficulty.

1 a The list shows the average gestation times, to the nearest day, of some animals. Show this data using a suitable stem-and-leaf diagram.

Animal	Gestation time (days)
Common opossum	13
Marine turtle	55
Grass lizard	42
Emperor penguin	63
House mouse	19
Royal albatross	79
Australian skink	30
Falcon	29
Hawk	44
Swan	30

Animal	Gestation time (days)
Python	61
Thrush	14
Wren	16
Spiny lizard	63
Alligator	61
Dog	63
Finch	12
Ostrich	42
Pheasant	22

b This data shows the track times in minutes and seconds on some of Luciano's CDs. Show the data using a suitable stem-and-leaf diagram.

3.21 2.29 2.25 2.49 2.57 3.30 3.19 2.25 3.34 2.45 2.44 3.34

3.19 3.30 2.10 3.00 2.44 2.25 2.54 3.43 2.22 2.54 2.55 2.24

3.35 2.10 3.55 3.07 2.54 2.08 3.22 3.33 2.43 3.50 2.22 2.57

2 This information gives the waiting times at J A Williams' dental practice.

Waiting time (minutes)	Frequency
$0 \leqslant t < 3$	9
$3 \leqslant t < 6$	17
$6 \leqslant t < 9$	12
$9 \leqslant t < 12$	6
$12 \leqslant t < 15$	2
$15 \leqslant t < 18$	1

a Draw a cumulative frequency diagram for this data.

Use your diagram to find:

b how many patients waited less than four minutes

c how many patients waited more than ten minutes.

d Use your diagram to estimate the values of the median and the interquartile range.

3 a Draw a box plot to represent this data:

1 3 5 6 8 10 11 12 14 15 15

b Draw a box plot to represent the information given in question **2**.

4 BROLLIES Я US have issued their sales figures for the past two years. (Figures are to the nearest £1000.)

	Year 1	Year 2
Jan	210	200
Feb	204	190
Mar	165	158
Apr	144	140
May	126	130
Jun	96	110
Jul	72	104
Aug	51	76
Sep	153	150
Oct	155	162
Nov	178	183
Dec	195	199

Draw a time series of these figures and include:

a a six-point moving average graph

b a twelve-point moving average graph.

c What conclusions do you draw from the graphs for the periods:

i January–April in Year 2

ii June–August in Year 2?

5 The distribution of ages of passengers on a train is given in the table.

Age range (years)	Frequency
$0 \leqslant a < 5$	11
$5 \leqslant a < 10$	32
$10 \leqslant a < 15$	28
$15 \leqslant a < 20$	35
$20 \leqslant a < 30$	45
$30 \leqslant a < 50$	58
$50 \leqslant a < 65$	115
$65 \leqslant a < 90$	76

a Draw a histogram to illustrate the data.

b How many people were on the train?

c Which age range made up the biggest proportion of passengers?

Try a real past exam question to test your knowledge:

6 The table summarises the distance thrown in the discus event by 20 boys during a school sports day.

Distance, x (m)	Number of boys
$0 < x \leqslant 5$	1
$5 < x \leqslant 10$	0
$10 < x \leqslant 20$	9
$20 < x \leqslant 30$	5
$30 < x \leqslant 35$	4
$35 < x \leqslant 40$	1

a Draw a histogram to represent this data.

The distances thrown in the discus event by 20 girls are represented by the histogram below.

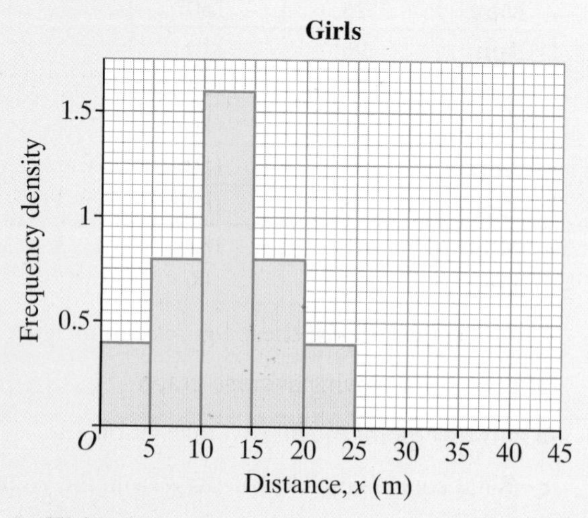

b Write down two comparisons between the distances thrown by the boys and the girls.

Spec B, Higher Paper, Nov 04

3 Scatter graphs

OBJECTIVES

D

Examiners would normally expect students who get a D grade to be able to:

Draw a scatter graph by plotting points on a graph

Interpret the scatter graph

Draw a line of best fit on the scatter graph

C

Examiners would normally expect students who get a C grade also to be able to:

Interpret the line of best fit

Identify the type and strength of the correlation

What you should already know ...

- Use coordinates to plot points on a graph
- Draw graphs including labelling axes and giving a title

VOCABULARY

Coordinates – a system used to identify a point; an x-coordinate and a y-coordinate give the horizontal and vertical positions

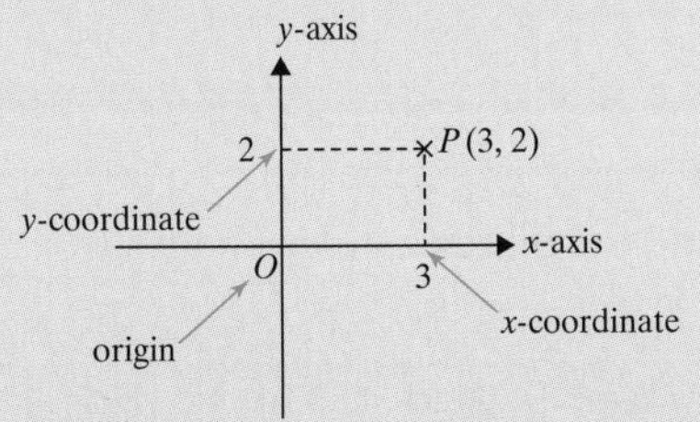

Correlation – a measure of the relationship between two sets of data; correlation is measured in terms of type and strength

Strength of correlation

The strength of correlation is an indication of how close the points lie to a straight line (perfect correlation)

Strong correlation

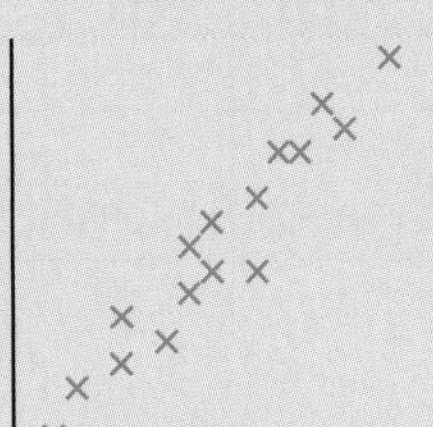

Weak correlation

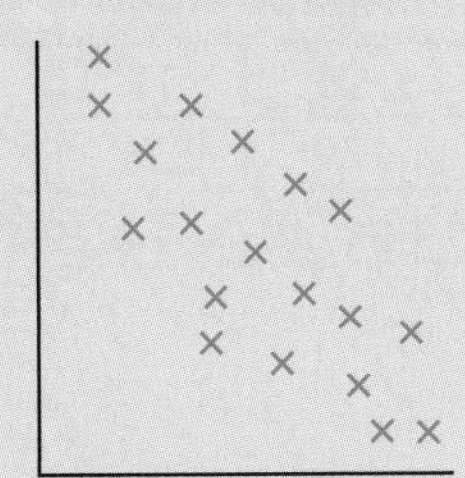

Correlation is usually described in terms of strong correlation, weak correlation or no correlation

Type of correlation

Positive correlation

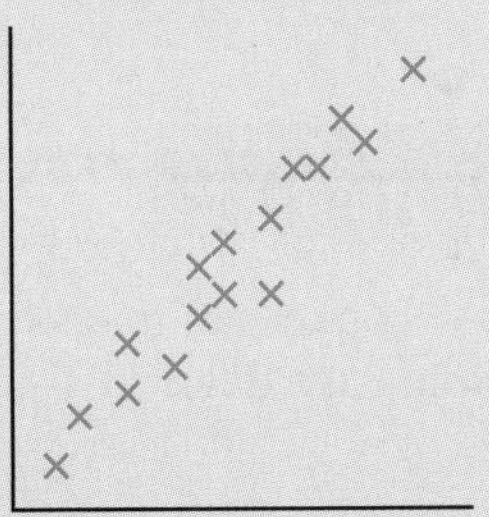

In positive correlation an increase in one set of variables results in an increase in the other set of variables

Negative correlation

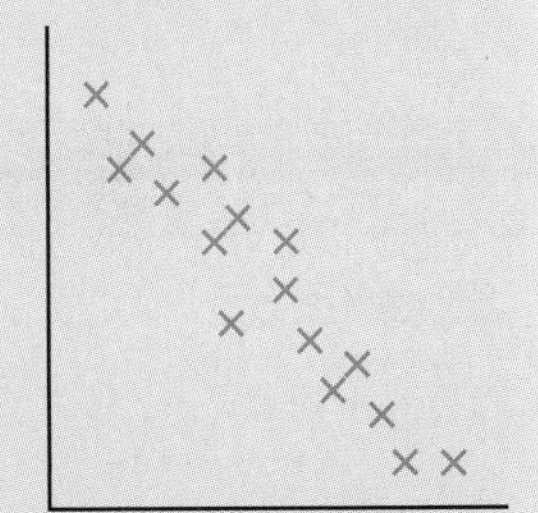

In negative correlation an increase in one set of variables results in a decrease in the other set of variables

Zero or no correlation

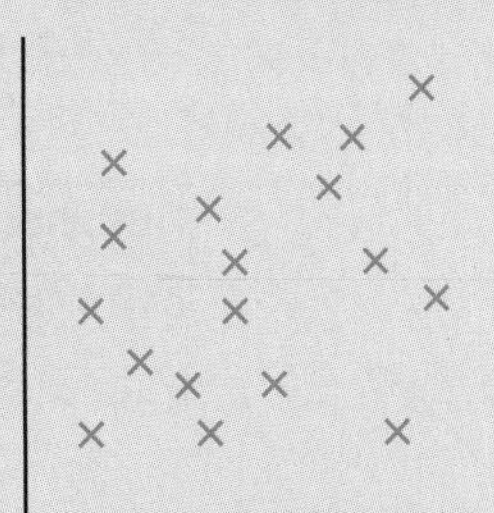

Zero or no correlation is where there is no obvious relationship between the two sets of data

Line of best fit – a line drawn to represent the relationship between two sets of data. Ideally it should only be drawn where the correlation is strong, for example,

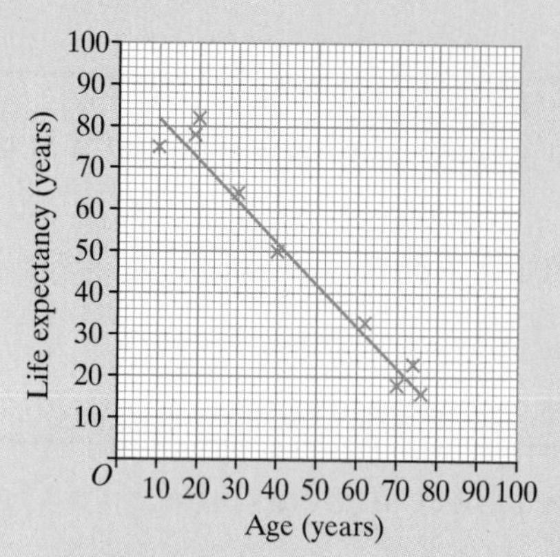

Scatter graph – a graph used to show the relationship between two sets of variables, for example, temperature and ice cream sales

Outlier – a value that does not fit the general trend, for example,

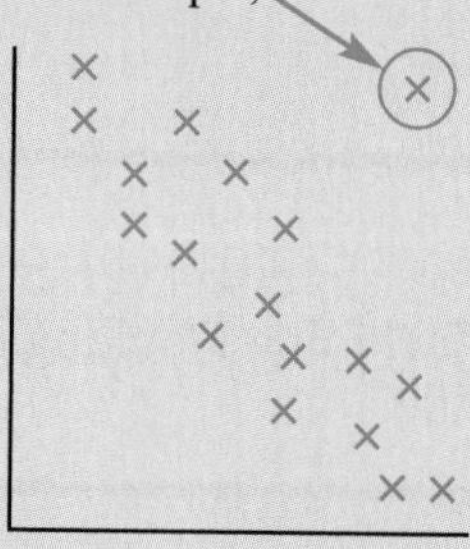

Learn 1 Scatter graphs

Example:

A shopkeeper notes the temperature and the number of ice creams sold each day.

	Sun	Mon	Tue	Wed	Thu	Fri	Sat
Temperature (°C)	20	26	17	24	30	15	18
Ice cream sales	35	39	27	36	45	25	32

Plot this information on a scatter graph and describe the correlation between temperature and ice cream sales.

The information can be plotted as a series of coordinate pairs.

	Sun	Mon	Tue	Wed	Thu	Fri	Sat
Temperature (°C)	20	26	17	24	30	15	18
Ice cream sales	35	39	27	36	45	25	32
	(20, 35)	(26, 39)	(17, 27)	(24, 36)	(30, 45)	(15, 25)	(18, 32)

Before drawing a scatter graph, you need to choose carefully the scales on the axes.

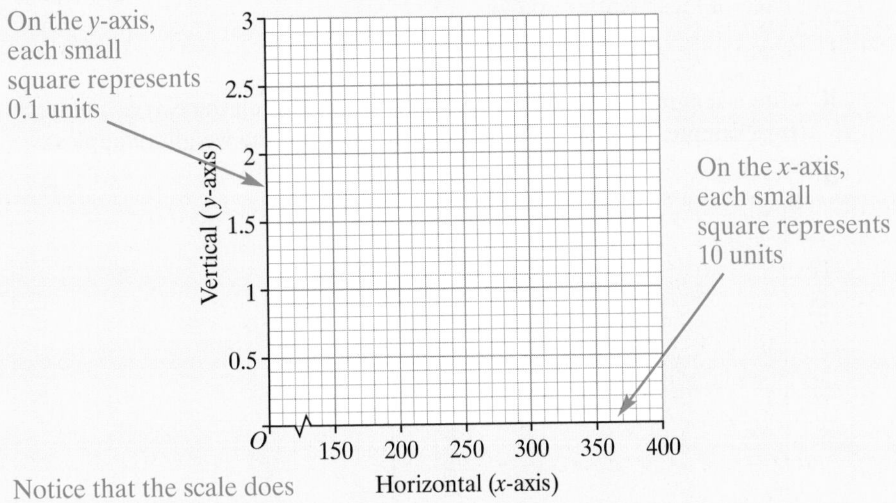

Notice that the scale does not always have to start at zero; a jagged line is often used to show the scale does not start at O

Temperature against ice cream sales

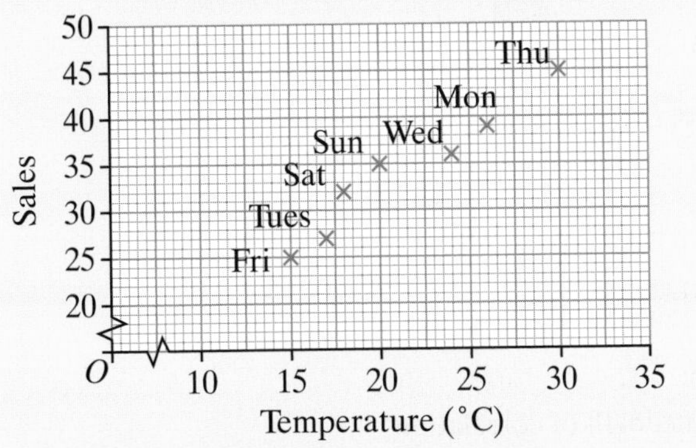

Plot the first variable on the horizontal axis

You can see from the graph that as the temperature rises, the sales of ice cream increase. There is a link between the temperature and the sales of ice cream.

Correlation measures the relationship between two sets of data

It is measured in terms of the **strength** and **type** of correlation

Therefore, there is **strong positive** correlation between the ice cream sales and the temperature.

Apply 1

1 For each of these scatter graphs:

i describe the type and strength of correlation

ii write a sentence explaining the relationship between the two sets of data (for example, the higher the rainfall, the heavier the weight of apples).

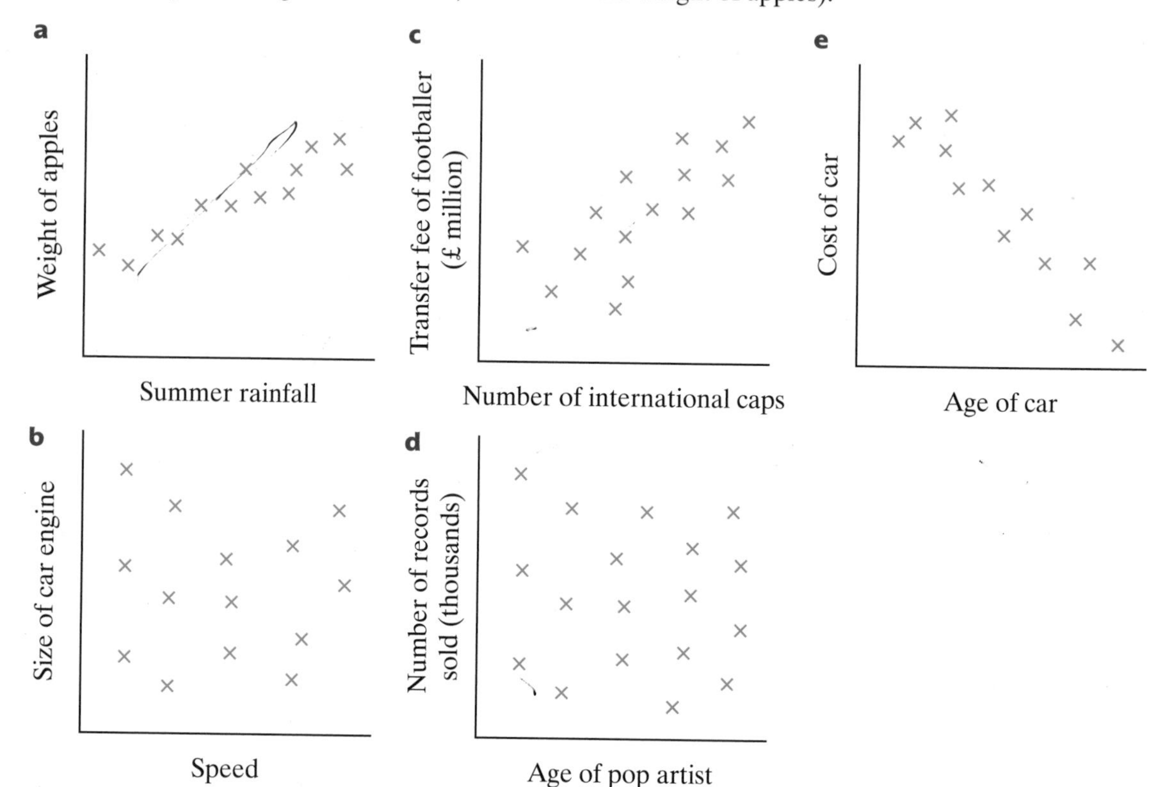

2 For each of the following:

i describe the type and strength of correlation

ii write a sentence explaining the relationship between the two sets of data.

a The hours of sunshine and the income from hiring deckchairs.

b The number of cars on one particular road and the average speed of cars.

c The distance travelled and the money spent on petrol.

d The height of a person and the number of children in his or her family.

3 The table shows the ages and arm spans of seven students in a school.

Age (years)	16	13	13	10	18	10	15
Arm span (inches)	62	57	59	57	64	55	61

a Draw a scatter graph of the results.

b Describe the type and strength of correlation.

c Write a sentence explaining the relationship between the two sets of data.

4 The table shows the ages and second-hand values of seven cars.

Age of car (years)	2	1	4	7	10	9	8
Value of car (£)	4200	4700	2800	1900	400	1100	2100

a Draw a scatter graph of the results.

b Describe the type and strength of correlation.

c Write a sentence explaining the relationship between the two sets of data.

5 The table shows the daily rainfall and the number of sunbeds sold at a resort on the south coast.

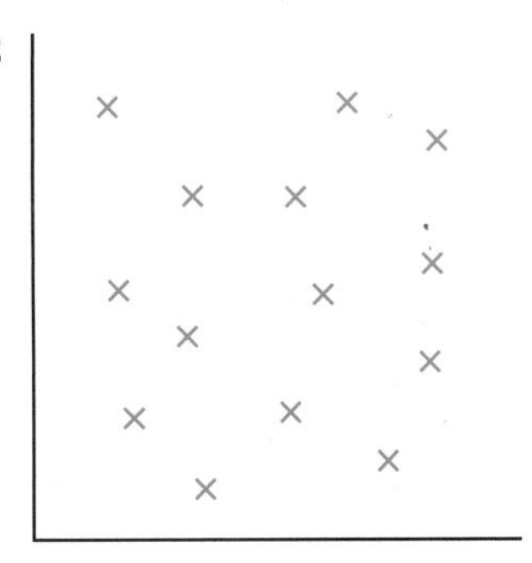

Amount of rainfall (mm)	0	1	2	5	6	9	11
Number of sunbeds sold	380	320	340	210	220	110	60

a Draw a scatter graph of the results.

b Describe the type and strength of correlation.

c Write a sentence explaining the relationship between the two sets of data.

6 For each graph, write down two variables that might fit the relationship.

a

b

c

7 The scatter graph shows the ages and shoe sizes of a group of people.

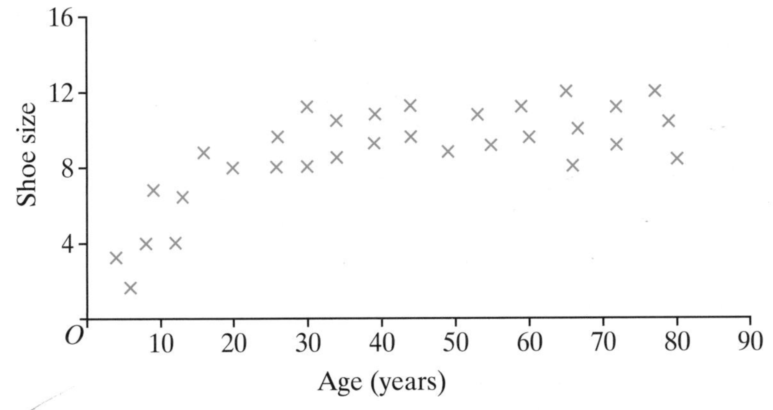

a Describe the correlation.

b Give a reason for your answer.

8 Get Real!

Steve is investigating the fat content and the calorie values of food at his local fast-food restaurant.
He collects the following information.

	Fat (g)	Calories
Hamburger	9	260
Cheeseburger	12	310
Chicken nuggets	24	420
Fish sandwich	18	400
Medium fries	16	350
Medium cola	0	210
Milkshake	26	1100
Breakfast	46	730

a Draw a scatter graph of the results.

b Describe the type and strength of correlation.

c Does the relationship hold for all the different foods?
Give a reason for your answer.

Explore

- Undertake some research of your own into the fat content and calorie values for food in a local restaurant
- You can find this information from the restaurant itself or on the internet

Investigate further

Learn 2 Lines of best fit

Example: Use the graph to draw a line of best fit and estimate the likely sales of ice cream for a temperature of 28°C.

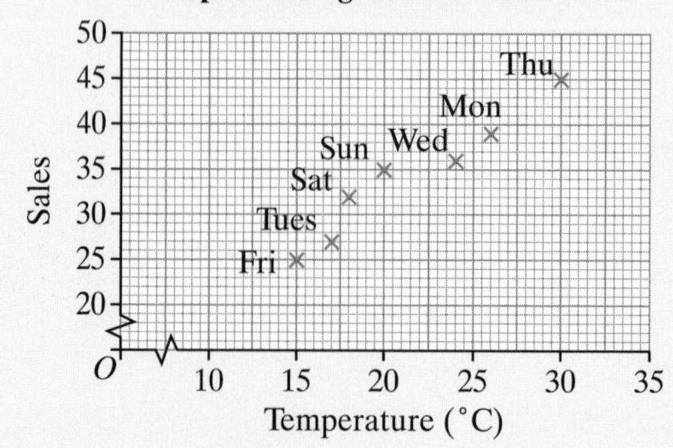

Drawing the line of best fit, you can use the graph to estimate the likely sales of ice creams for a temperature of 28°C.

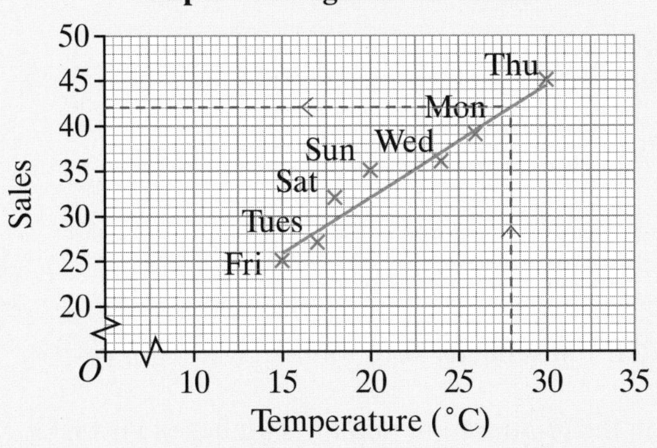

For some graphs, a straight line is not possible and a curve of best fit may be appropriate

From the graph, the number of ice cream sales is 42.

However, care should be taken when making use of such methods. For example, you will notice that the sales of ice creams were slightly higher on Saturday and Sunday.

You might also use your line of best fit to estimate the temperature given the number of ice cream sales – but it is not likely that you would need to do this.

Apply 2

1 The table shows the ages and arm spans of seven students in a school.

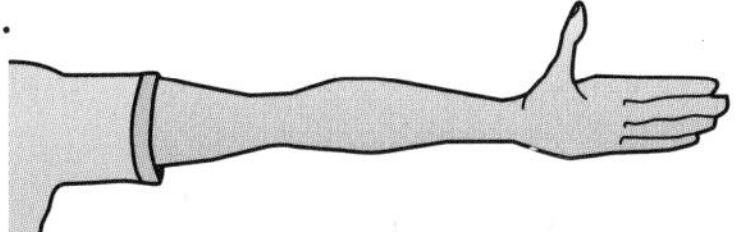

Age (years)	16	13	13	10	18	10	15
Arm span (inches)	62	57	59	57	64	55	61

a Draw a line of best fit on the scatter graph you drew in Apply **1**, question **3**.

b Use your line of best fit to estimate:

i the arm span of an 11-year-old student

ii the age of a student with an arm span of 61 inches.

2 The table shows the ages and second-hand values of seven cars.

Age of car (years)	2	1	4	7	10	9	8
Value of car (£)	4200	4700	2800	1900	400	1100	2100

a Draw a line of best fit on the scatter graph you drew in Apply **1**, question **4**.

b Use your line of best fit to estimate:

i the value of a car if it is 7.5 years old

ii the age of a car if its value is £3700.

3 The table shows the daily rainfall and the number of sunbeds sold at a resort on the south coast.

Amount of rainfall (mm)	0	1	2	5	6	9	11
Number of sunbeds sold	380	320	340	210	220	110	60

a Draw a line of best fit on the scatter graph you drew in Apply **1**, question **5**.

b Use your line of best fit to estimate:

i the number of sunbeds sold for 4 mm of rainfall

ii the amount of rainfall if 100 sunbeds were sold.

4 Sally collects information on the temperature and the number of visitors to a museum.

Temperature (°C)	15	25	16	18	19	22	24	23	17	20	26	20
Number of visitors	720	180	160	620	510	400	310	670	720	530	180	420

a Draw a scatter graph and a line of best fit.

b Use your line of best fit to estimate:

i the number of visitors if the temperature is 21°C

ii the temperature if 350 people visit the museum.

c Sally is sure that two of the data pairs are incorrect. Identify these two pairs on your graph.

5 The graph shows the line of best fit for the relationship between house prices in 2000 and house prices in 2006.

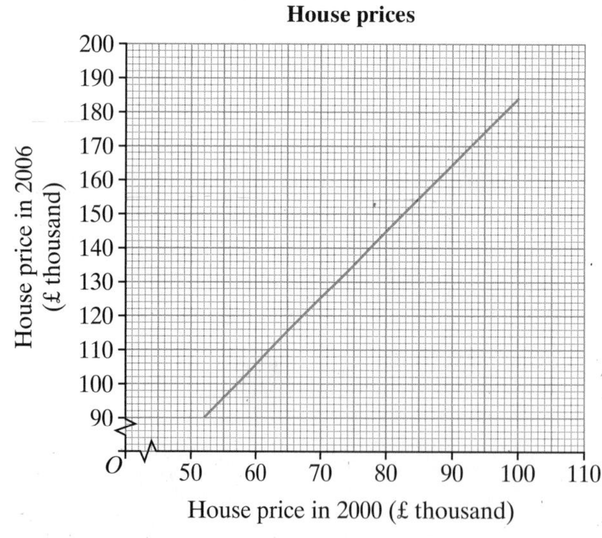

a Copy and complete the table, giving estimates for the missing values.

House price in 2000 (£ thousand)	70	75	90	100		
House price in 2006 (£ thousand)	125				105	155

b Andrea says her house price was £80 000 in 2000 and £170 000 in 2006. Is she correct? Give a reason for your answer.

c Find an estimate for the 2006 price of a house priced £55 000 in 2000.

d Find an estimate for the 2000 price of a house priced £180 000 in 2006.

6 Jenny collects information on the maximum speed and engine size of various motorbikes.
Her results are shown in the table.

Engine size (cc)	50	250	350	270	400	440	600	800	950	900	1200	1000
Speed (kph)	70	120	140	150	180	190	220	250	270	260	270	240

a Draw a scatter graph.

b What do you notice about the correlation between speed and engine size?

c Draw a line of best fit and use this to estimate:

i the engine size if your speed is 170 kph

ii the engine size if your speed is 250 kph.

d Which of these results is likely to be the most reliable?
Give a reason for your answer.

7 Readings of two variables, A and B, are shown in the table.

A	1	2	3	4	5	6	7	0.8	2.1	3.2	3.9	5.1	6.2	7.1
B	1.8	8.8	20	33	48	73	95	2	9	18	31	49	72	98

a Draw a scatter graph.

b What can you say about the correlation between the two sets of data?

c Draw a curve of best fit and use this to estimate:

i the value of B if $A = 2.5$

ii the value of A if $B = 64$.

d Express the relationship between A and B.

Explore

- Investigate the cost and age of cars – you may wish to use the internet
- What relationship do you notice?
  Are there any exceptions?

Investigate further

Explore

- Investigate house prices and the number of bedrooms – you may wish to use a local newspaper
- What relationship do you notice? Are there any exceptions?

Investigate further

Explore

- Investigate body lengths – for example, your arm span and your height are closely related
- Collect data from your friends and family to see what relationships you can find

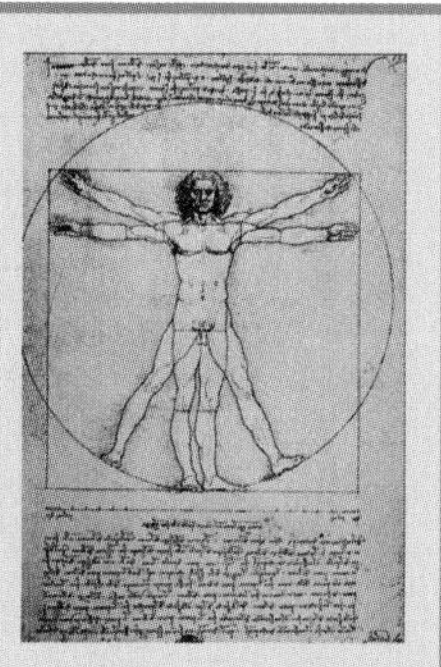

Investigate further

Scatter graphs

ASSESS

The following exercise tests your understanding of this chapter, with the questions appearing in order of increasing difficulty.

1 The table shows the marks of eight students in English and mathematics.

Student	1	2	3	4	5	6	7	8
English	25	35	28	30	36	44	15	21
Mathematics	27	40	29	32	41	48	17	20

Draw a scatter graph and comment on the relationship between the marks in the two subjects.

2 Mr Metcalf, the maths teacher, told his class they had a test in a week's time. He also asked them to record how many hours of TV they watched during the week before the test.
When he had marked the test he showed the class a scatter graph of the data in the table.

Student	1	2	3	4	5	6	7	8	9	10
TV watched (hours)	4	7	9	10	13	14	15	20	21	25
Test mark	92	90	74	30	74	66	95	38	35	30

a Draw a scatter graph and comment on the relationship between the marks in the test and the amount of TV watched.

b Two students do not seem to 'fit the trend'.
Identify the students and give a possible reason for their results.

3 The tables show the relationship between the areas (in thousands of km^2) of some countries and their populations (in millions), all to 2 s.f.

	Area (km^2)	**Population (millions)**
Monaco	0.0020	0.030
Malta	0.32	0.40
Jersey	0.12	0.090
Netherlands	42	16
UK	250	60
Germany	360	83
Italy	300	58
Switzerland	41	7.3
Andorra	0.47	0.068
Denmark	43	5.4

	Area (km^2)	**Population (millions)**
France	550	60
Austria	84	8.2
Turkey	780	67
Greece	130	11
Spain	500	40
Ireland	70	3.9
Latvia	65	2.4
Sweden	450	8.9
Norway	320	4.5
Iceland	100	0.28

Draw a scatter graph of this data and comment on the graph.

4 a Draw a suitable scatter graph to illustrate this data, which shows the relationship between the distances jumped in long jump trials and the leg lengths of the jumpers.

Leg length (cm)	71	73	74	75	76	79	82
Distance jumped (m)	3.2	3.1	3.3	4.1	3.9	4	4.8

b Draw a line of best fit on the graph.

c Use your line of best fit to estimate:

i the leg length of an athlete who jumped a distance of 3.5 m

ii the distance jumped by an athlete with a leg length of 85 cm.

d Explain why one of these estimates is more reliable than the other.

Try a real past exam question to test your knowledge:

5 The scatter graph shows the height and trunk diameter of each of eight trees.

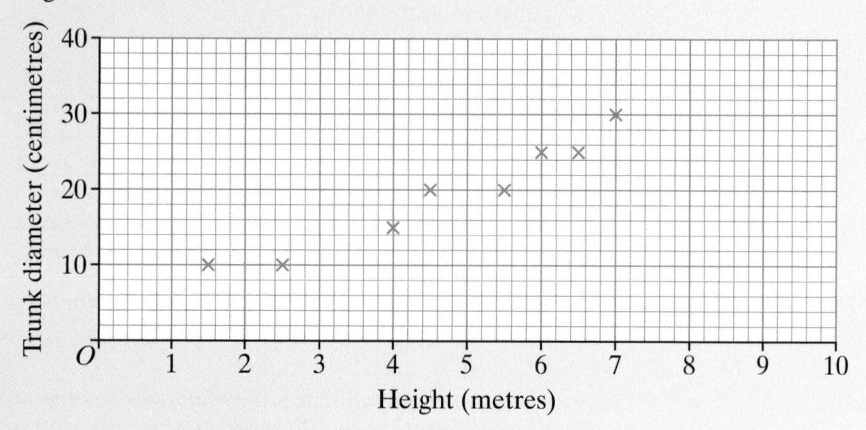

a What is the height of the tallest tree?

b Draw a line of best fit through the points on the scatter graph.

c Describe the relationship shown in the scatter graph.

d i Estimate the height of a tree with trunk diameter 35 centimetres.

ii Comment on the reliability of your estimate.

Spec B, Int Paper 1, Mar 02

4 Collecting data

OBJECTIVES

D **Examiners would normally expect students who get a D grade to be able to:**

Classify and know the difference between various types of data

Use a variety of different sampling methods

Design and use data collection sheets and questionnaires

C **Examiners would normally expect students who get a C grade also to be able to:**

Identify possible sources of bias in the design and use of data collection sheets and questionnaires

A **Examiners would normally expect students who get an A grade also to be able to:**

Use stratified sampling methods

What you should already know ...

- Counting and, in particular, counting in fives (for tally charts)
- Design and use tally charts for discrete and grouped data
- Design and use two-way tables for discrete and grouped data

VOCABULARY

Tally chart – a useful way to organise the raw data; the chart can be used to answer questions about the data, for example,

Number of pets	Tally
0	卌 IIII
1	卌 卌 II
2	卌 II
3	III
4	II

The tallies are grouped into five so that

IIII = 4

卌 = 5

卌 I = 6

This makes the tallies easier to read

Two-way table – a combination of two sets of data presented in a table form, for example,

	Men	Women
Left-handed	7	6
Right-handed	20	17

Quantitative data – data that can be counted or measured using numbers, for example, number of pets, height, weight, temperature, age, shoe size, etc.

Qualitative or **categorical data** – data that cannot be measured using numbers, for example, type of pet, car colour, taste, people's opinions/feelings, etc.

Discrete data – data that can only be counted and take certain values, for example, number of cars (you can have 3 cars or 4 cars but nothing in between, so $3\frac{1}{2}$ cars is not possible)

Continuous data – data that can be measured and take any value; length, weight and temperature are all examples of continuous data

Survey – a way of collecting data; there are a variety of ways of doing this, including face-to-face, or via telephone, e-mail or post using questionnaires

Respondent – the person who answers the questionnaire

Direct observation – collecting data first-hand, for example, counting cars at a motorway junction or observing someone shopping

Primary data – data that you collect yourself; this is new data and is usually gathered for the purpose of a task or project (including GCSE coursework)

Secondary data – data that someone else has collected; this might include data in books, newspapers, magazines, etc. or data that has been loaded onto a database

Data collection sheets – these are used to record the responses to the different questions on a questionnaire; they can also be used with computers to load data onto a database

Pilot survey – a small-scale survey to check for any unforeseen problems with the main survey

Convenience or **opportunity sampling** – a survey that is conducted using the first people who come along, or those who are convenient to sample (such as friends and family)

Random sampling – this requires each member of the population to be assigned a number; the numbers are then chosen at random

Systematic sampling – this is similar to random sampling except that it involves every nth member of the population; the number n is chosen by dividing the population size by the sample size

Quota sampling – this method involves choosing a sample with certain characteristics, for example, select 20 adult men, 20 adult women, 10 teenage girls and 10 teenage boys to conduct a survey about shopping habits

Cluster sampling – this is useful where the population is large and it is possible to split the population into smaller groups or clusters

Stratified sampling – this involves dividing the population into a series of groups or 'strata' and ensuring that the sample is representative of the population as a whole, for example, if the population has twice as many boys as girls, then the sample should have twice as many boys as girls

Learn 1 Collecting data

Examples:

a Categorise each of the following:

i Weight of whales in an aquarium
ii Favourite make of car of students in the sixth form
iii Number of bottles of lemonade sold at a store each day
iv Best pop group in the charts.

	Quantitative	Qualitative	Continuous	Discrete
i Weight of whales in an aquarium	✓		✓	
ii Favourite make of car of students in the sixth form		✓		
iii Number of bottles of lemonade sold at a store each day	✓			✓
iv Best pop group in the charts		✓		

b Write down one advantage and one disadvantage of the following methods of collecting data.

i Personal surveys.
ii Postal surveys.
iii Direct observations.

i This is the most common method of collecting data and involves an interviewer asking questions of the interviewee. This method is sometimes called a face-to-face interview.

Advantages
- The interviewer can ask more complex questions and explain them if necessary.
- The interviewer is likely to be more consistent when they record the responses.
- The interviewee is more likely to answer the questions than with postal or e-mail surveys.

Disadvantages
- This method of interviewing takes a lot of time and can be expensive.
- The interviewer can influence the answers and this may cause bias.
- The interviewee is more likely to lie or to refuse to answer a question.

The telephone survey is a special case of the personal survey with similar advantages and disadvantages

ii Postal surveys make use of mailing lists (or the electoral register) and involve people being selected and sent a questionnaire.

Advantages
- The interviewees can take their time answering and give more thought to the answer.
- The possibility of interviewer bias is avoided.
- The cost of a postal survey is usually lower.

Disadvantages
- Postal surveys suffer from low response rates which may cause bias.
- The process can take a long time to get questionnaires out and await their return.
- Different people might interpret questions in different ways when giving their answers.

The e-mail survey is a special case of the postal survey and an increasingly popular method for collecting data

iii Direct observation, as the name implies, means observing the situation directly. Direct observation might include, for example, counting cars at a motorway junction or observing someone to see what shopping they buy. It can take place over a short or a long period of time.

Advantages
- Direct observation is a reliable method which allows observations to take place in the interviewees 'own environment'.

Disadvantages
- For some experiments, the interviewee may react differently because they are being observed.
- This method takes a lot of time and can be expensive.

c Write down three requirements of a good questionnaire.

- **Appropriate** to the survey being carried out and not asking unnecessary questions.
  For example, asking someone for their address may not be appropriate for most questionnaires ... unless it is a survey on where people live
- **Unbiased** so that they do not lead the respondent to give a particular answer.
  For example, asking the question 'Do you believe we should have a new shopping centre?' is a biased (leading) question
- **Unambiguous** so that they are clear and straightforward to the respondent.
  For example, asking the question 'Do you agree or disagree that we should have a new shopping centre?' is not very clear

Apply 1

1 For each of the following say whether the data is quantitative or qualitative.

a The number of people at a cricket test match.

b The weights of newborn babies.

c How many cars a garage sells.

d Peoples' opinions of the latest Hollywood blockbuster.

e The best dog at Crufts.

f The time it takes to run the London Marathon.

g The colour of baked beans.

h How well your favourite football team played in their last match.

i The number of text messages received in a day.

2 For each of the following say whether the data is discrete or continuous.

a The number of votes for a party at a general election.

b The ages of students in your class.

c The number of beans in a tin.

d The weight of rubbish each household produces each week.

e How many people watch the 9 o'clock news.

f How long it takes to walk to school.

g The number of sheep Farmer Angus has.

h The weights of Farmer Angus' sheep.

i The heights of Year 10 students in your school.

3 Connect the following to their proper description.
The first one has been done for you.

Lengths of fish caught in a competition	**Quantitative and discrete**
Number of goals scored by a football team	
Ages of teachers at a school	
Favourite colours in the tutor group	**Qualitative**
Number of sweets in a bag	
Favourite type of music	
Person's foot length	**Quantitative and continuous**
Person's shoe size	
Cost of stamps	
Best player on the Wales rugby team	

4 For each of the following questions, identify whether the information requested is:

- quantitative and discrete
- quantitative and continuous
- qualitative.

Questionnaire

Age (years)	
Gender (M/F)	
Height (cm)	
Hand span (cm)	
Arm span (cm)	
Foot length (cm)	
Eye colour	
No. of brothers	
No. of sisters	
House number	
Favourite pet	

5 The following questions are taken from different surveys.
Write down one criticism of each question.
Rewrite the question in a more suitable form.

a How many hours of TV do you watch each week?

Less than 1 hour ☐ More than 1 hour ☐

b What is your favourite football team?

Real Madrid ☐ Luton Town ☐

c How do you spend your leisure time? (You can only tick one box.)

Doing homework ☐ Playing sport ☐ Reading ☐

Computer games ☐ On the Internet ☐ Sleeping ☐

d You do like football, don't you?

Yes ☐ No ☐

e How much do you earn each year?

Less than £10 000 ☐ £10 000 to £20 000 ☐ More than £20 000 ☐

f How often do you go to the cinema?

Rarely ☐ Sometimes ☐ Often ☐

g Do you or do you not travel by taxi?

Yes ☐ No ☐

6 Peter and Paul are writing questions for their coursework task.

a This is one question from Peter's questionnaire:

> Skateboarding is an excellent pastime. Don't you agree?
> Tick one of the boxes.
>
> Strongly agree ☐ Agree ☐ Don't know ☐

Write down two criticisms of Peter's question.

b These questions are from Paul's questionnaire:

> Do you buy CDs? ☐ Yes ☐ No
>
> If yes, how many CDs do you buy on average each month?
>
> ☐ 2 or less ☐ 3 or 4 ☐ 5 or 6 ☐ more than 6

Write down two reasons why these are good questions.

7 Write down a definition for:

  a quantitative data
  b qualitative data
  c continuous data
  d discrete data.

8 Write down two advantages of undertaking a pilot survey.

9 Write down five things that make a good questionnaire.

Explore

Here are some investigations for you to consider.

- A company slogan states 'In tests 9 out of every 10 cats prefer it'
- Investigate ways in which the company could have arrived at this answer
- You might consider carrying out your own survey
- How might you get such results?

- In a supermarket poll of 100 people 85 said they had at least two cars!
- Is this correct?
- Investigate ways in which the company could have arrived at this answer
- How might you get such results?

Investigate further

Learn 2 Sampling methods

Examples:

You are given a list of 500 students (200 boys and 300 girls) and wish to choose a sample of 50.

Explain how you would use the following sampling methods.

a Convenience sampling
b Random sampling
c Systematic sampling
d Quota sampling
e Cluster sampling
f Stratified sampling

a Convenience sampling or opportunity sampling means that you just take the first people who come along or those who are convenient to sample. The likelihood is that you choose the first 50 students that you meet or otherwise choose 50 students from among your friends.

b Random numbers can be taken from random number tables or generated by a calculator using the RAN or RND button. Assign each student a number between 0 and 499 and generate random numbers to choose 50 students.

c Systematic sampling is similar to random sampling except that you take every nth member of the population. The value n is found by dividing the population size by the sample size giving $\frac{500}{50} = 10$ so that every 10th student is chosen from the population. A random number is used to start so that the number 4 would suggest taking the 4th, 14th, 24th, 34th... students.

d Quota sampling is popular in market research and involves choosing a sample with certain characteristics (the choice of who to ask is left to the interviewer). The likely requirement is that you are asked to sample 20 boys and 30 girls.

e Cluster sampling is useful where the population is large and it is possible to split the population into smaller groups or clusters. The most obvious choice is to consider tutor groups as clusters and sample the whole of two tutor groups, although this is not likely to result in exactly 50 students.

f Stratified sampling involves dividing the population into a series of groups or 'strata' and ensuring that the sample is representative of the population as a whole. For example, if the population has twice as many boys as girls then the sample should have twice as many boys as girls.

You choose 20 boys and 30 girls randomly using one of the above sampling methods. Stratified sampling would also take account of different year groups and these would need to be divided in a similar ratio.

g The table shows the number of students in each year group of a school.

Year	7	8	9	10	11
Number	200	200	240	220	140

A stratified sample ensures that the sample is representative of the population as a whole. To take a stratified sample, you need to appreciate that 200 students out of 1000 students are from Year 7, so that $\frac{200}{1000}$ of the sample should be from Year 7

Completing this information for each year group:

Year	7	8	9	10	11
Number	200	200	240	220	140
Number	$\frac{200}{1000}$	$\frac{200}{1000}$	$\frac{240}{1000}$	$\frac{220}{1000}$	$\frac{140}{1000}$
For a sample size of 50	$\frac{200}{1000} \times 50$ = 10 students	$\frac{200}{1000} \times 50$ = 10 students	$\frac{200}{1000} \times 50$ = 12 students	$\frac{200}{1000} \times 50$ = 11 students	$\frac{200}{1000} \times 50$ = 7 students

If the required sample size is 50, then $\frac{200}{1000} \times 50$ will be from Year 7, that is 10 students from Year 7

When completing this information for each section, remember to check that the totals for each year group add up to the sample size of 50

Apply 2

1 The following table shows the ordered ages for 100 people in a London shopping centre.

22	22	22	22	22	22	23	23	24	24
24	24	24	24	24	24	24	25	25	25
25	25	25	25	25	26	26	26	26	26
26	26	26	26	26	27	27	27	27	27
28	28	28	28	28	28	28	28	29	29
29	29	29	30	31	31	31	31	31	31
32	32	33	33	33	33	33	33	34	34
34	34	34	34	34	34	34	34	35	35
35	35	35	36	36	36	36	36	36	36
37	38	40	40	42	42	44	47	48	50

a Take a random sample of 20 people using the RAN or RND button on your calculator.
Find the mean of this sample.

b Take a systematic sample of 20 people using every fifth number.
Find the mean of this sample.

c Take a cluster sample of the first 20 people in the table.
Find the mean of this sample.

d Which sample do you think is most representative of the 100 people?
Give a reason for your answer.

2 Which sampling method is most appropriate for each of the following surveys? Give a reason for your answer.

a The amount of pocket money students receive in different year groups at your school.

b The favourite programmes of 15 to 19 year olds.

c The average life expectancy of people around the world.

d The favourite holiday destinations of people in the sixth form at your school.

e The opinions of two hundred 25 to 39 year olds on their favourite soap opera.

f Information on voting intentions at a general election.

3 Consider each of the following surveys and say whether the sample is representative. Give a reason for your answer.

a Aiden is trying to find out what students in the school think about school dinners. He decides to use cluster sampling, asking the first 30 Year 7s as they leave the dining room.

b Betty wants to find out the most popular names for newborn babies – she goes to the Internet, finds the relevant website and takes the first 100 names on the list.

c Cameron wants to find out if all premiership footballers think that having overseas players is a good thing. He decides to take a 10% sample and ask all the players at Chelsea and Arsenal.

d Davina wishes to check how many people travel on the underground. She telephones 100 people at home in the evening and asks them if they have travelled on the underground that week.

e Eric undertakes a convenience sample of 20 friends to see if they wear glasses. He says that 45% of students at his college wear glasses.

4 Explain the difference between a random sample and a systematic sample.

5 Find two articles containing around 200 words each.
Find which article contains longer words.

a Choose an appropriate sampling method and give a reason for your choice.

b Choose an appropriate sample size and give a reason for your choice.

6 A college wishes to undertake a survey on the eating habits of its students.
Explain how you would take:

a a random sample of 100 students

b a systematic sample of 100 students

c a stratified sample of 100 students.

7 The table shows the number of students in each year group of a college.

Year	10	11	12	13
Number	300	300	220	180

Explain how you would take a stratified sample of size 50.

8 The table shows the number of people employed in a department store.

Occupation	Management	Sales	Security	Office
Number	10	130	25	35

Explain how you would take a stratified sample of size:

a 40　　　　**b** 50

9 Write one advantage and one disadvantage of each of these sampling methods.

a Convenience sampling

b Random sampling

c Systematic sampling

d Quota sampling

e Stratified sampling

Explore

- Design your own questionnaire and undertake a survey
- You should consider a hypothesis and write some suitable questions
- Design a questionnaire and undertake a pilot survey
- Amend your questionnaire and explain any improvements
- Consider a suitable sampling method – explain why you have chosen this method
- Carry out the survey and comment on your results

Investigate further

Collecting data

ASSESS

The following exercise tests your understanding of this chapter, with the questions appearing in order of increasing difficulty.

1 For each of the following, state whether the data is quantitative or qualitative.

a The heights of the people watching a tennis match.

b The colours of cars in a car park.

c The sweetness of different orange juices.

d The lengths of pencils in a pencil case.

e The numbers of students in different classes.

f The numbers of leaves on different types of trees.

g The ages of people at a night club.

h The musical abilities of students in a class.

2 For each of the following, state whether the data is continuous or discrete.

a The heights of the people watching a tennis match.

b The times taken by athletes to complete a race.

c The numbers of sweets in a sample of packets.

d The lengths of pencils in a pencil case.

e The numbers of students in different classes.

f The numbers of leaves on different types of trees.

g The shoe sizes of people at a party.

h The ages of people at a night club.

3 **a** Criticise each of the following questionnaire questions.

i How many hours of television have you watched in the last 2 months?

ii Do you or do you not watch news programmes?

b Criticise each of the following questionnaire questions and suggest alternatives to find out the required information.

i What do you think about our new improved fruit juice?

ii How much do you earn?

iii Do you or do you not agree with the new bypass?

iv Would you prefer to sit in a non-smoking area?

v How often do you have a shower?

4 **a** Explain why each of the following may not produce a truly representative sample.

i Selecting people at random outside a supermarket.

ii Selecting every 10th name from an electoral register starting with M.

iii Selecting names from a telephone directory.

b In a telephone poll conducted one morning, 20 people were asked whether they regularly used a bus to get to work. Give three reasons why this sample might not be truly representative.

5 **a** Explain how you could use the electoral register to obtain a simple random sample.

b A college wishes to undertake a survey on the part-time employment of its students. Explain how you would take:

i a random sample of 100 students

ii a stratified sample of 100 students.

6 The table shows the number of people employed in a factory.

Occupation	Management	Office	Sales	Shop floor
Number	10	15	30	145

a Explain why a random sample of the employees might not be suitable.

b Explain how you would take a stratified sample of size 40.

c Explain how you would take a stratified sample of size 50.

5 Probability

OBJECTIVES

D

Examiners would normally expect students who get a D grade to be able to:

Use a two-way table to find a probability

Understand mutually exclusive outcomes

Use the fact that the probabilities of mutually exclusive outcomes add up to 1

C

Examiners would normally expect students who get a C grade also to be able to:

Use probability to estimate outcomes for a population

B

Examiners would normally expect students who get a B grade also to be able to:

Use relative frequency to find probabilities

Complete a tree diagram

A

Examiners would normally expect students who get an A grade also to be able to:

Understand dependent and independent outcomes

Use tree diagrams to find probabilities of successive independent events

A*

Examiners would normally expect students who get an A* grade also to be able to:

Draw tree diagrams and use them to find probabilities of successive dependent events

What you should already know ...

- How to use a probability scale
- The difference between experimental and theoretical probability
- How to calculate simple probabilities
- How to find outcomes systematically

VOCABULARY

Probability – a value between 0 and 1 (which can be expressed as a fraction, decimal or percentage) that gives the likelihood of an event

Outcome – the result of an experiment, for example, when you toss a coin, the outcome is a head or a tail

Certain – an outcome with probability 1, for example, the sun rising and setting

Likely – an outcome with a probability greater than $\frac{1}{2}$, for example, rain falling in November in the UK

Unlikely – an outcome with a probability less than $\frac{1}{2}$, for example, snow falling in August in the UK

Impossible – an outcome with a probability 0, for example, the sun turning green

Evens – probability $\frac{1}{2}$, for example, there is an even chance of getting a head or a tail when you toss a coin

Theoretical probability – probability based on equally likely outcomes, for example, it suggests you will get 5 heads and 5 tails if you toss a coin 10 times

Experimental probability or **relative frequency** – this is found by experiment, for example, if you get 6 heads and 4 tails, the experimental probability would be $\frac{6}{10}$ or 0.6 for getting a head

Random – a choice made when all outcomes are equally likely, for example, picking a raffle ticket from a box with your eyes shut

Mutually exclusive events – these cannot both happen in the same experiment, for example, getting a head and a tail on one toss of a coin

Two-way table or **sample space diagram** – table used to show all the possible outcomes of an experiment, for example, all the outcomes of tossing a coin and throwing a dice

		Dice					
		1	2	3	4	5	6
Coin	Head	H1	H2	H3	H4	H5	H6
	Tail	T1	T2	T3	T4	T5	T6

Independent events – events are independent when the outcome of one does not affect the outcome of the other, for example, tossing a coin and drawing a card from a pack

Dependent events – events are dependent when the outcome of one affects the outcome of the other, for example, taking two successive balls from a box without replacing the first one

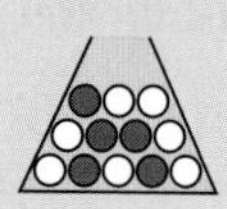
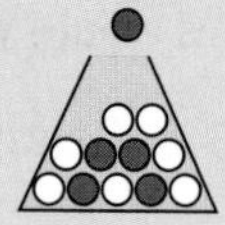
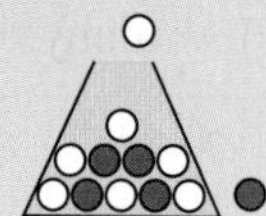

Learn 1 Experimental and theoretical probability

Examples:

The experimental and theoretical results of throwing a dice 600 times are shown in the table below.

	Outcome					
	1	2	3	4	5	6
Theoretical results	100	100	100	100	100	100
Experimental results	92	107	103	99	97	102

The probability of getting a six from one throw of a dice is $\frac{1}{6}$

This is a theoretical probability which leads to the assumption that there will be 100 sixes in 600 throws of the dice

In practice, it is unlikely that there will be exactly 100 of each number

a Use the table to calculate:

i the theoretical probability of getting a 6.

The theoretical probability is $\frac{100}{600} = \frac{1}{6}$

The theoretical probability cancels down to $\frac{1}{6}$

ii the experimental probability of getting a 6.

The experimental probability is $\frac{102}{600}$

The experimental probability is also called the relative frequency

b Do you think the dice is biased? Give a reason for your answer.
The dice is not biased as the theoretical and experimental frequencies are close to each other.

An unbiased dice is usually called a fair dice

Some events cannot be predicted theoretically, for example, the probability that there will be 12 sunny days in August; the data for weather in August over the past five years could be used to find an experimental probability for this outcome

Apply 1

Apart from question 8, this is a non-calculator exercise.

1 A coin is tossed 300 times.

How many times would you expect to get a head?

2 The probability of getting a faulty light bulb is 0.01

How many faulty light bulbs would you expect to find in a batch of 500?

3 A spinner with five equal divisions labelled A, B, C, D, E is spun 100 times.

a How many times would you expect it to land on A?

b How many times would you expect it to land on D?

c Julie spins the spinner 10 times and gets 2 As, 3 Bs, no Cs and 5 Ds.

She says this shows the spinner is biased. Is she correct?

Explain your answer.

4 The probability that an inhabitant of Random Island has red hair is $\frac{3}{20}$

There are 4263 people living on Random Island.

How many of them would you expect to have red hair?

5 The table shows the frequency distribution after taking a letter at random from the word ASSESS 60 times.

	Results from drawing a letter 60 times		
	A	**E**	**S**
Frequency	13	10	37

a What is the relative frequency of getting the letter S?

b How does this compare with the theoretical probability of getting the letter S?

c If this experiment is repeated 600 times, how many times would you expect to get the letter E?

6 Kali has a spinner with coloured sections of equal size.
She wants to know the probability that her spinner lands on pink.
She spins it 100 times and calculates the relative frequency of pink after every 10 spins.
Her results are shown on the graph.

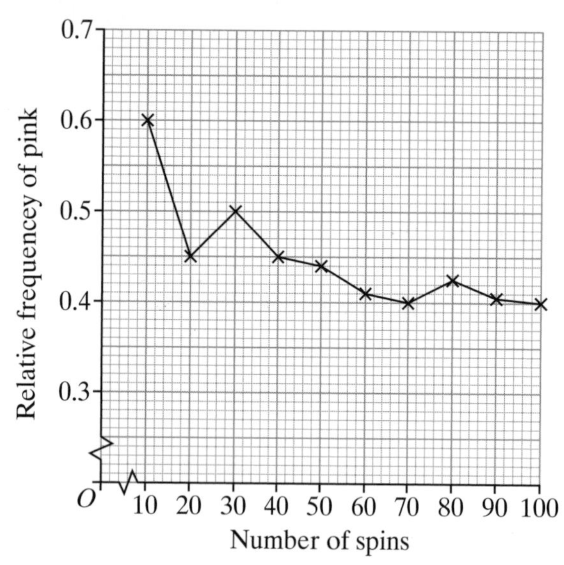

a Use the graph to calculate the number of times that the spinner landed on pink:

i after the first 10 spins

ii after the first 50 spins.

b From the graph, estimate the probability of the spinner landing on pink.

c Kali's results confirm that her spinner is fair.
The spinner has five equal sections.

i How many sections are pink?

ii Kali spins the spinner two more times.
What is the theoretical probability that the spinner lands on pink both times?

7 A spinner is made from a regular octagon.
Each of the eight sections of the spinner has one of the letters P, Q, R, S.
It is spun 240 times and the results are shown in the table.

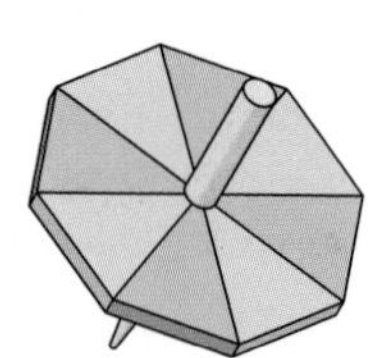

	Results from 240 spins			
	P	Q	R	S
Frequency	34	62	88	56

How many times does each letter appear on the spinner?

8 Emma has three dice – one is red, one is blue and one is green.
The table shows the frequency distributions after throwing each dice 600 times.

	Results from 600 throws of dice					
	1	**2**	**3**	**4**	**5**	**6**
Red dice	111	108	89	98	96	98
Blue dice	54	65	83	121	129	148
Green dice	88	99	106	96	102	109

a One of the dice appears to be biased. Which one is this?
What suggests that it is biased?

b From the two fair dice, what is the relative frequency of throwing a 5?
How does this compare with the theoretical probability?

c From the two fair dice, what is the relative frequency of throwing a number greater than 3?
How does this compare with the theoretical probability?

9 Sam has some counters of assorted colours in a bag.
He asks his friends to pick a counter with their eyes shut.
He wants to find the probability of getting a red counter.
After every 10 attempts, he finds the relative frequency of red and the results are shown on the graph.

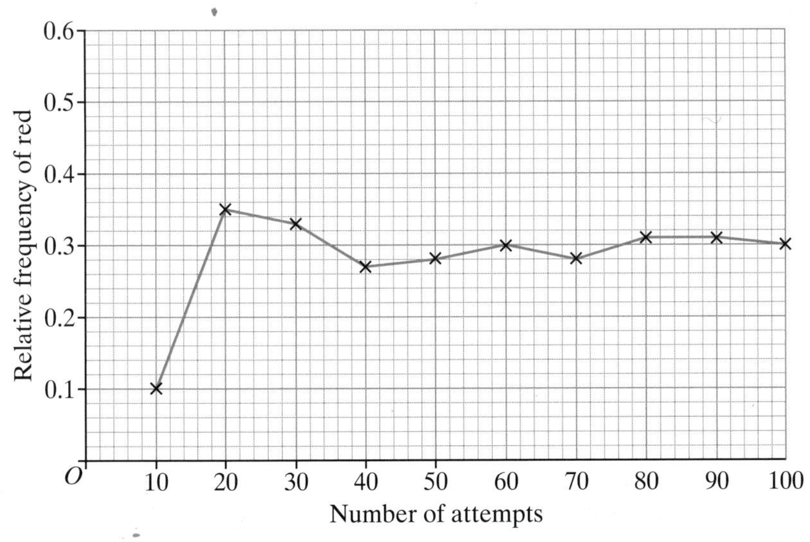

a How many red counters had been picked after 10 attempts?

b How many red counters had been picked after 50 attempts?

c From the graph, estimate the probability of picking a red counter.

d If there are 12 red counters, how many counters are in the bag altogether?

Explore

- Throw two dice together and write down the sum of their scores
- Repeat this until you have recorded 30 totals
- Which total appears the most?
- Which total appears the least?

Investigate further

Learn 2 Combined events

Example:

Mollie tosses a coin and picks a letter at random from the word SPACE.
Draw a table to show this information.
What is the probability that she gets a head and a vowel?

	S	P	A	C	E
Head	Head, S	Head, P	Head, A	Head, C	Head, E
Tail	Tail, S	Tail, P	Tail, A	Tail, C	Tail, E

These results have been set out systematically in a two-way table, so that none are left out

There are 10 different possible outcomes.

There are 2 outcomes with a head and a vowel.

P(head and vowel) = $\frac{2}{10} = \frac{1}{5}$

Remember to cancel down where possible

Apply 2

1 The Aces, Kings, Queens and Jacks are taken from a full pack of cards to form a new, smaller pack. A card is then drawn at random from this pack.

	A	K	Q	J
Clubs ♣	A♣	K♣		
Hearts ♥	A♥			
Diamonds ♦	A♦			
Spades ♠				

a Copy and complete the table of outcomes (two-way table) and use it to find the probability of drawing:

b the King of diamonds

c a black Queen

d a red Jack *or* a red Ace

e a black card

f the King *or* Queen of clubs.

2 Alan has five tiles and three cards as shown in the diagram.

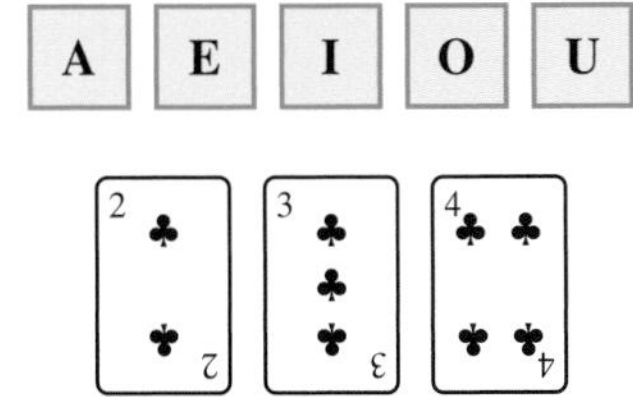

He picks up one tile and one card at random.

a Draw a table of outcomes.

Find the probability that Alan picks:

b tile A and the 3 of clubs

c tile A and a card that is *not* the 3 of clubs

d a tile that is *not* A and a card with an even number.

3 Two dice are thrown and their scores are added together.

a Copy and complete the two-way table.

	1	**2**	**3**	**4**	**5**	**6**
1	2		4			
2		4				
3	4					
4						
5						
6						

Use your two-way table to find:

b the probability of getting a total of 4

c the probability of getting a total of 14

d the probability of getting a total of 7

e the probability of getting a total which is at least 5.

4 A box contains 4 green counters and 1 red counter.
Mo chooses one counter at random and Sean throws a dice.

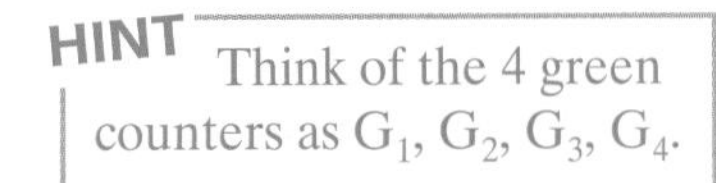

a Draw a two-way table.

b Find the probability that Mo gets a red counter and Sean throws a 5.

c Find the probability that Mo gets a green counter and Sean throws a 6.

d Find the probability that Mo gets a red counter and Sean throws an even number.

5 Tim picks a card at random from a full pack.
Jo picks a marble at random from a bag containing 2 black marbles and 3 red marbles. Find the probability that:

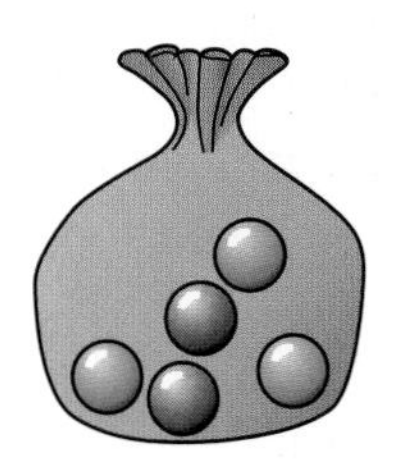

a Tim picks a diamond and Jo picks a black marble

b Tim's card is from a red suit and Jo's marble is also red

c Tim does not pick a diamond and Jo picks a red marble.

6 Two dice are thrown and the product of their scores is recorded.

a Copy and complete the two-way table.

	1	2	3	4	5	6
1	1					
2			6			
3						
4						
5					25	
6						

Use your two-way table to find:

b the probability of getting a product of 15

c the probability of getting a product of 12

d the probability of getting a product greater than 16

e the probability of getting a product that is an even number.

7 Lucy says that if you toss two coins, the possible outcomes are two heads, two tails or one of each. She therefore decides that the probability of getting two heads is $\frac{1}{3}$

Explain why Lucy is wrong.

8 Shaun tosses three coins.

a How many possible outcomes are there?

b What is the probability that he gets one head and two tails?

c What is the probability that he gets at least one head?

Explore

- How many possible outcomes are there from tossing 2 coins?
- How many possible outcomes are there from tossing 3 coins?
- How many possible outcomes are there from tossing 4 coins?
- How many possible outcomes are there from tossing 5 coins?

Investigate further

Learn 3 Mutually exclusive events

Examples:

a Are the outcomes 'throwing a 6' and 'throwing a 2' mutually exclusive?

If you throw a dice once, you may get a 6 and you may get a 2.

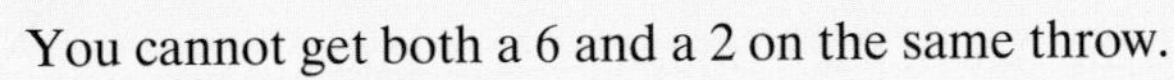

You cannot get both a 6 and a 2 on the same throw.

The outcome 'throwing a 6' and the outcome 'throwing a 2' are mutually exclusive.

b Are the outcomes 'drawing a diamond' and 'drawing a 6' mutually exclusive?

If you draw one card from a pack, you may get a diamond and you may get a 6.

You could get the 6 of diamonds, so the outcome 'drawing a diamond' and the outcome 'drawing a 6' are *not* mutually exclusive.

The sum of the probabilities of all the mutually exclusive events is 1

If event A and event B are mutually exclusive, then P(A or B) = P(A) + P(B).

c There are 7 white counters, 8 black counters and 5 yellow counters in a bag. One is drawn at random.
Find the probability of getting:

i a white counter

ii a black counter

iii a yellow counter

iv a black or a yellow counter

v a counter that is not white.

Drawing a white counter, drawing a black counter and drawing a yellow counter are mutually exclusive events.

Remember to cancel down where possible

i P(white) $= \frac{7}{20}$

ii P(black) $= \frac{8}{20} = \frac{2}{5}$

iii P(yellow) $= \frac{5}{20} = \frac{1}{4}$

Adding these probabilities we get $\frac{7}{20} + \frac{8}{20} + \frac{5}{20} = \frac{20}{20} = 1$
Use this as a check on your work

iv P(black or yellow) = P(black) + P(yellow) $= \frac{8}{20} + \frac{5}{20} = \frac{13}{20}$

v P(not white) $= 1 - \frac{7}{20} = \frac{13}{20}$

This is the same as P(black or yellow)

Apply 3

1 Which of these are mutually exclusive events?

a Getting a six and an odd number on a throw of a dice.

b Getting a four and an even number on a throw of a dice.

c Picking a heart and the Ace of clubs from a pack of cards.

d Picking the Queen of diamonds and the Jack of diamonds from a pack of cards.

e Picking a club and a King from a pack of cards.

2 The probability that Nicola will come top in the maths exam is 0.8
What is the probability that she will not come top?

3 The probability that Andy will fail his driving test is $\frac{4}{15}$
What is the probability that he will pass?

4 The probability that Random United will win their next match is 0.3
The probability that they will draw is 0.6
What is the probability that they will lose?

5 The probability that Steve will score more than 50% in his history test is 0.7
Jenny says, 'That means there is a probability of 0.3 that he will get less than 50%'.
Explain why she is wrong.

6 The table shows the probabilities that a student at Random High School will choose certain lunch menus.

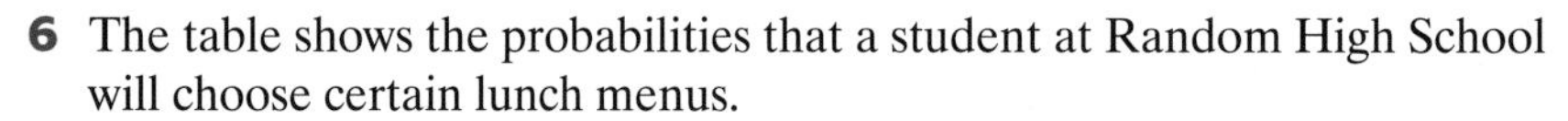

Salad	Pizza	Hotpot	Other
0.2	...	0.1	0.3

What is the probability that the student will choose pizza?

7 There are three possible transport routes from Canary Wharf to Waterloo: a waterbus, the Docklands Light Railway (DLR) or the Jubilee Line.
The probability that Andy uses the waterbus is 0.16
The probability that he uses the DLR is 0.4
Helen says this means there is a probability of 0.8 that he uses the Jubilee Line.
What mistake has she made? What is the correct probability?

8 The table shows the probabilities that the next car going past the school will be a particular colour.

Silver	Red	Black	White	Other
0.34	0.15	0.14	...	0.3

a What is the probability that the next car will be white?

b What is the probability that the next car will be either red or black?

9 Errol goes out to buy a new sweatshirt.
The probability that he buys it from Supershirts is $\frac{9}{20}$
The probability that he buys it from BestSweats is $\frac{2}{5}$
What is the probability that he buys it from one of these stores?

10 The probability that Zoe will move to Rugby is 0.4
The probability that she will move to Coventry is 0.2
The probability that she will move to Leicester is 0.3

a What is the probability that Zoe will move to either Rugby or Leicester?

b What is the probability that she will *not* move to Coventry?

c What is the probability that she will not move to any of these three towns?

11 Jayne has a collection of 50 old teapots.
27 of them have rose patterns. Some of them are covered in blue willow pattern. Some of them have striped patterns. The rest have plain colours.
One teapot is chosen at random.

The probability that it will have a striped pattern is 0.08

a How many of Jayne's teapots have a striped pattern?

The probability that it will have a blue willow pattern is 0.22

b How many of Jayne's teapots have plain colours?

12 Matt has 25 books on his bookshelf. Some of them are history textbooks, some are science fiction, five are biographies and two are dictionaries.
He has put his library card in one of the books but cannot remember which one.

a Find the probability that the library card is in either a biography or a dictionary.

The probability that it is in a history textbook is 0.4.

b How many science fiction books are on the shelf?

Learn 4 Independent events

Example:

Kathy tosses a coin and Liam draws a card from a pack at random.
What is the probability that Kathy throws a head and Liam draws an Ace?

Draw a two-way table:

Liam

		A	2	3	4	5	6	7	8	9	10	J	Q	K
Kathy	**H**	HA	H2	H3	H4	H5	H6	H7	H8	H9	H10	HJ	HQ	HK
	T	TA	T2	T3	T4	T5	T6	T7	T8	T9	T10	TJ	TQ	TK

The probability that Kathy tosses a head and Liam draws an Ace is $\frac{1}{26}$

Alternatively P(head) $= \frac{1}{2}$

P(ace) $= \frac{1}{13}$

P(head and ace) $= \frac{1}{2} \times \frac{1}{13} = \frac{1}{26}$

If the events are independent then the probability that they will both happen is found by multiplying the probabilities together

If events A and B are independent then P(A and B) = P(A) × P(B)

Apply 4

1 Anna throws a dice and Bob draws a card at random.
What is the probability that:

a Anna gets a 6 and Bob gets a King

b Anna gets an odd number and Bob gets a diamond

c Anna gets a number greater than 2 and Bob gets a red 5?

2 The probability that Jo will come top in maths is 0.3
The probability that she will come top in French is 0.4
What is the probability that she comes top in both subjects?

3 The probability that Steve has porridge for breakfast is $\frac{1}{4}$
The probability that he catches a bus to work is $\frac{3}{5}$
What is the probability that Steve has porridge and catches the bus to work?

4 The probability that Paul washes his car on Saturday is $\frac{1}{6}$
The probability that he rents a video is $\frac{2}{9}$
What is the probability that Paul does not wash his car but does rent a video?

5 Lorraine has a packet of crisps and a fruit juice for lunch every day.
The probability that she has roast chicken crisps is 0.2
The probability that she has orange juice is 0.4
What is the probability that she has roast chicken crisps and orange juice?

6 Two dice are thrown together.
What is the probability that the combined score is 2?

7 Three dice are thrown together.
What is the probability that the combined score is 18?

8 The probability that Lucy will go to Spain for her holiday is 0.55
The probability that she will go on holiday in July is 0.44
Marie says that means Lucy is 99% certain to go to Spain in July.
Explain why Marie is wrong.

9 Dee, Karen and Matt are all taking their driving tests next week.
The probability that Dee will pass is 0.7
The probability that Karen will pass is 0.9
The probability that Matt will pass is 0.6
The results of their tests are independent of each other.

a What is the probability that all three of them will pass?

b What is the probability that Dee and Karen will pass and Matt will fail?

c What is the probability that all three of them will fail?

10 The probability that Random United will win their next match is $\frac{7}{12}$
The probability that AQA Rovers will win their next match is $\frac{3}{7}$
They are not playing each other in their next match.

a What is the probability that both teams will win?

b What is the probability that neither team will win their match?

Learn 5 Tree diagrams

Example:

A bag contains 4 black counters and 5 white counters.
Two counters are taken at random from the bag.
Draw a tree diagram to show the information.

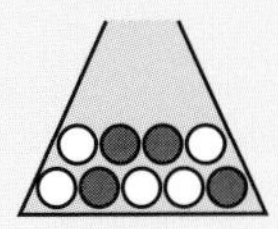

The answer depends on whether the first counter is replaced (independent events) or not replaced (dependent events).

'Replaced' tells you that the second event is independent of the first event

The possibilities are shown separately below.

i With replacement (independent events)
One counter is taken at random from the bag and then replaced.
A second counter is then taken at random from the bag.

The two events are *independent* because the colour of the first counter does not affect the outcome when the second counter is taken

The tree diagram is shown below.

A tree diagram is useful for calculating probabilities
The probabilities are written on the branches of the tree

1st counter **2nd counter**

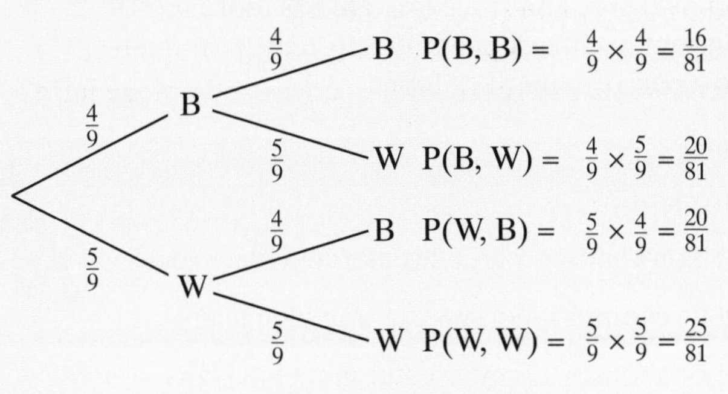

The first counter was replaced so it did not affect the outcome (or the probabilities) when the second counter was taken out

The two branches always add up to one, because they show all the possible outcomes

ii Without replacement (dependent events)
One counter is taken at random from the bag and not replaced.
A second counter is then taken at random from the bag.

The two events are *dependent* because the colour of the first counter does affect the outcome when the second counter is taken

'Not replaced' tells you the second event is dependent on the first event

1st counter **2nd counter**

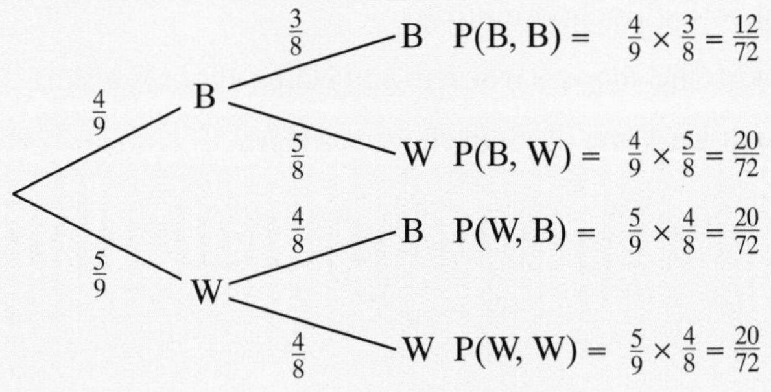

The first counter was not replaced so it affected the outcome when the second counter was taken out

If the first counter is black, there are only 3 black counters left and only 8 counters in the bag. So the probability of getting a second black counter becomes $\frac{3}{8}$

There are still 5 white counters, so the probability of getting a white counter becomes $\frac{5}{8}$

Apply 5

1 A box contains 3 red pencils and 7 green pencils. A pencil is taken from the box and then replaced. A second pencil is then taken from the box.

a Copy and complete the tree diagram.

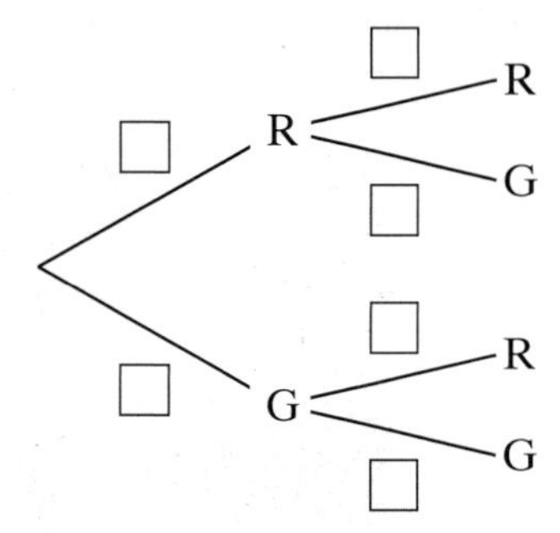

Use the tree diagram to find:

b the probability that both pencils are red

c the probability that both pencils are green

d the probability that one pencil is red and one is green.

2 Sam has five sweaters – two grey and three tan. He has four pairs of jeans – three blue pairs and one black pair. On Saturday morning he chooses a sweater and a pair of jeans at random.

a Draw a tree diagram.

Use the tree diagram to find:

b the probability that Sam chooses a grey sweater and black jeans

c the probability that he chooses a tan sweater and blue jeans

d the probability that he chooses a grey sweater and blue jeans.

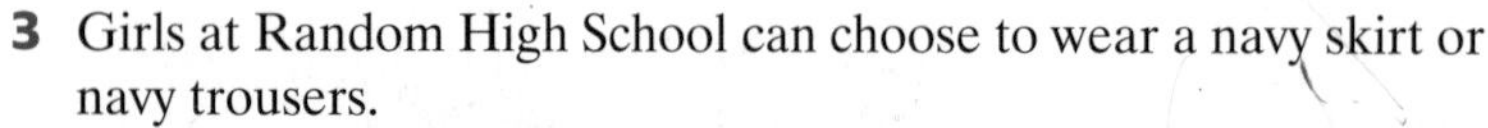

3 Girls at Random High School can choose to wear a navy skirt or navy trousers.
The probability that Mollie will choose a skirt is 0.4 and the probability that her friend Nadia will choose a skirt is 0.7

a Draw a tree diagram.

Use the tree diagram to find:

b the probability that both girls choose skirts

c the probability that Mollie chooses trousers and Nadia chooses a skirt

d the probability that at least one of them chooses a skirt.

4 The probability that Rajesh will oversleep is 0.2
If he oversleeps, the probability that he will be late for school is 0.8
If he does not oversleep, the probability that he will be late is 0.1

a Draw a tree diagram.

Use the tree diagram to find:

b the probability that Rajesh does not oversleep and is on time for school

c the probability that he is late for school.

5 The probability that George will play cricket after school is $\frac{2}{5}$
If he plays cricket, the probability that he will forget his homework is $\frac{5}{8}$
If he does not play cricket, the probability that he will forget his homework is $\frac{1}{4}$

a Draw a tree diagram.

Use the tree diagram to find:

b the probability that George plays cricket and forgets his homework

c the probability that he does not play cricket and remembers his homework

d the probability that he forgets his homework.

6 Amy has five pound coins and seven 20p pieces in her purse.
She takes out two coins at random. (Think of this as taking first one coin, then another, without replacement.)

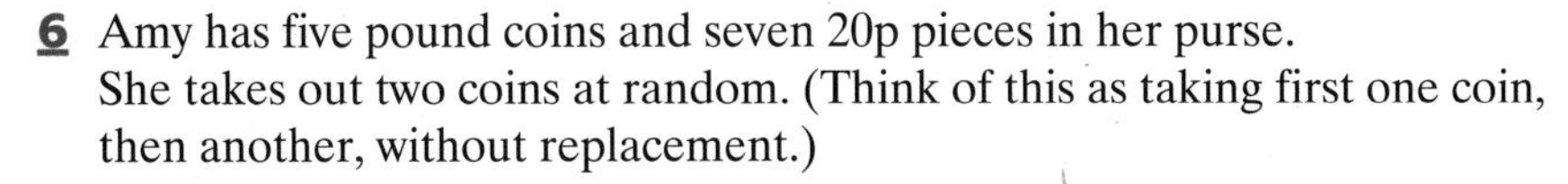

a Draw a tree diagram.

Use the tree diagram to find:

b the probability that Amy takes two 20p pieces

c the probability that she takes two pound coins

d the probability that she takes at least one pound coin.

7 Two cards are drawn at random from a full pack.

a Draw a tree diagram.

Use the tree diagram to find:

b the probability that both cards are Aces

c the probability that neither card is an Ace

d the probability that just one card is an Ace.

8 Dean has 8 black socks, 6 grey socks and 4 red socks in his drawer.
He takes out one sock and then another.

a Copy and complete the tree diagram.

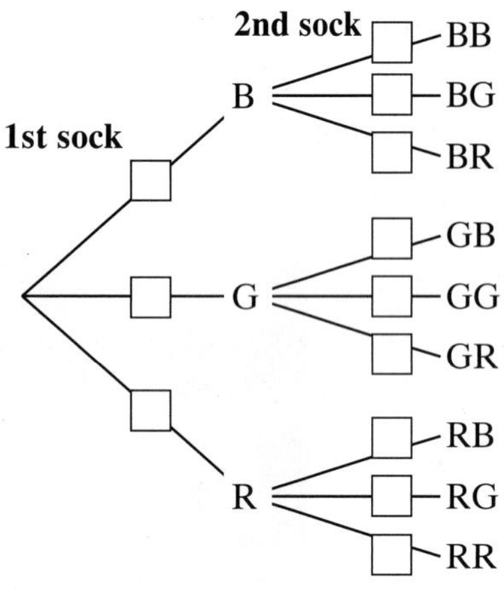

Use the tree diagram to find:

b the probability that Dean has picked two black socks

c the probability that Dean has picked one grey and one red sock

d the probability that Dean has picked a pair of matching socks.

9 The probability that Kylie will pass her driving test is $\frac{1}{3}$
This probability does not change, however many times she takes the test.
Find the probability that:

a Kylie passes her test at the second attempt

b Kylie passes her test at the fifth attempt

c Kylie passes her test at the nth attempt.

Probability

ASSESS

The following exercise tests your understanding of this chapter, with the questions appearing in order of increasing difficulty.

1 Two pentagonal spinners, each with the numbers 1 to 5, are spun and their outcomes added together to give a score.

a Draw a two-way table for the two spinners.

b Use your table to find:

i the probability of a score of 4

ii the probability of a score of 5

iii the probability of a score of 9

iv the most likely score.

2 **a** Which of these pairs of events are mutually exclusive events?

i Throwing a 3 and an even number on a throw of a dice.

ii Throwing a 1 and an odd number on a throw of a dice.

iii Picking a spade and a club from a pack of cards.

iv Picking a diamond and a King from a pack of cards.

b The probability that a train arrives early is 0.09
The probability that it arrives late is 0.4
What is the probability that it arrives on time?

c The table shows the probabilities of selecting tickets from a bag. The tickets are coloured yellow, black or green and numbered 1, 2, 3 or 4.

	1	2	3	4
Yellow	$\frac{1}{20}$	$\frac{1}{16}$	$\frac{3}{40}$	$\frac{1}{8}$
Black	$\frac{1}{10}$	$\frac{3}{40}$	0	$\frac{3}{40}$
Green	0	$\frac{1}{8}$	$\frac{3}{16}$	$\frac{1}{8}$

A ticket is taken at random from the bag. Calculate the probability that:

i it is black and numbered 4

ii it is numbered 2

iii it is green

iv it is yellow or numbered 1.

3 Nazeem selected one pen from a box containing 3 red, 4 green and 1 blue pens and a second pen from another box containing 2 red and 1 green. Copy and complete the tree diagram to show the possible outcomes.

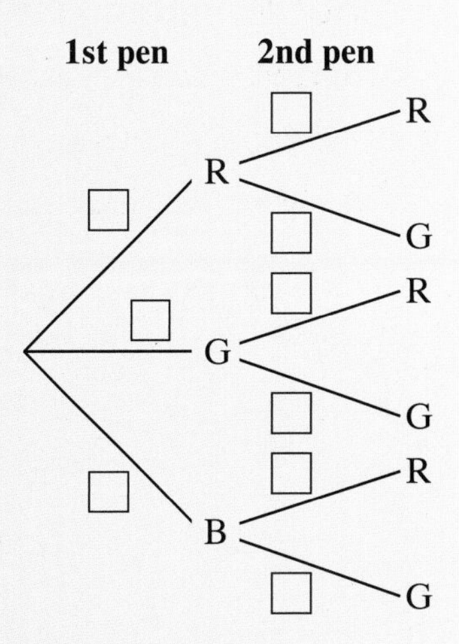

4 **a** Use your tree diagram from question **3** to find the probability of each possible outcome.

b Which outcome is the:

i least likely to occur **ii** most likely to occur?

c Nazeem actually selected two pens of the same colour. Which colour is she most likely to have selected?

d Find the probability that the two pens selected were:

i the same colour **ii** different colours.

5 In Class 11A at Thornes Comprehensive there are 21 girls and 9 boys. Everyone in the school has hair that can be classed as blonde, brown or black and the probability of having hair of that colouring is:

Blonde	0.2
Brown	0.3
Black	0.5

a Draw a tree diagram which shows all six possible outcomes for gender and hair colouring.

b One student is chosen at random from the class. Calculate the probability that this student is:

i a boy with brown hair

ii a girl with black hair

iii not blonde.

Glossary

Average – a single value that is used to represent a set of data

Back-to-back stem-and-leaf diagram – a stem-and-leaf diagram used to represent two sets of data

Number of minutes to complete a task

Leaf (units) Girls	Stem (tens)	Leaf (units) Boys
7 7 6 5 4 2 2	1	1 6 7 8 9
7 6 4 3 2 1	2	2 2 7 7 7 7 8 9
7 0	3	1 4 6

Key: 3 | 2 represents 23 minutes

Key: 3 | 4 represents 34 minutes

Box plot or **box and whisker plot** – used to show how the data is distributed

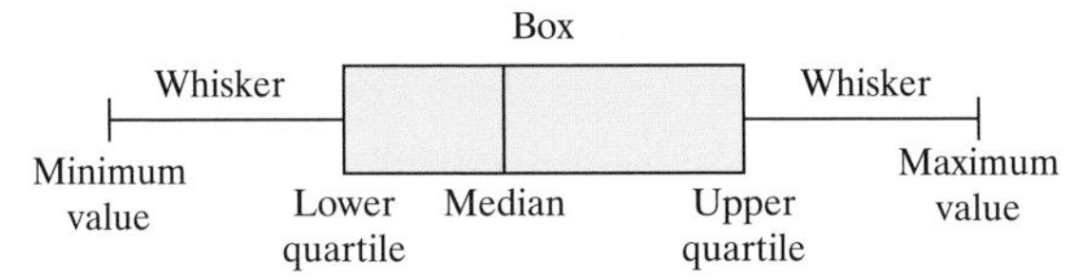

Categorical data – see qualitative data

Certain – an event with probability 1, for example, the sun rising and setting

Cluster sampling – this is useful where the population is large and it is possible to split the population into smaller groups or clusters

Continuous data – data that can be measured and take any value; length, weight and temperature are all examples of continuous data

Convenience or **opportunity sampling** – a survey that is conducted using the first people who come along, or those who are convenient to sample (such as friends and family)

Coordinates – a system used to identify a point; an x-coordinate and a y-coordinate give the horizontal and vertical positions

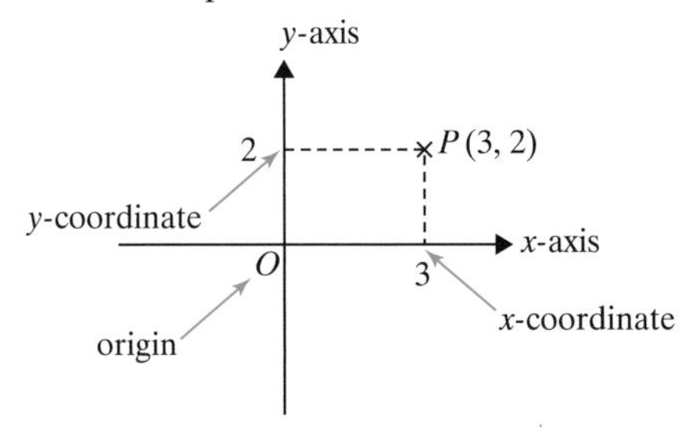

Correlation – a measure of the relationship between two sets of data; correlation is measured in terms of type and strength

Strength of correlation

The strength of correlation is an indication of how close the points lie to a straight line (perfect correlation)

Strong correlation

Weak correlation

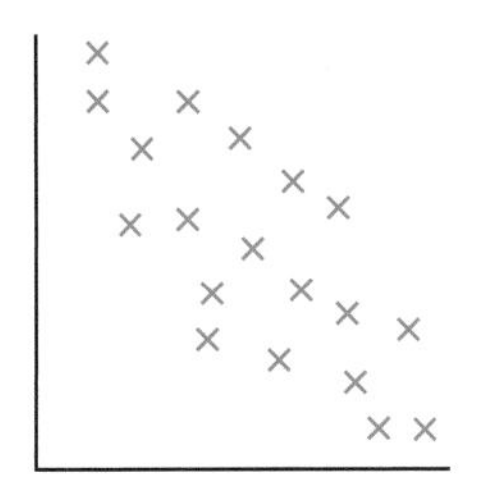

Correlation is usually described in terms of strong correlation, weak correlation or no correlation

Type of correlation

Positive correlation

In positive correlation an increase in one set of variables results in an increase in the other set of variables

Negative correlation

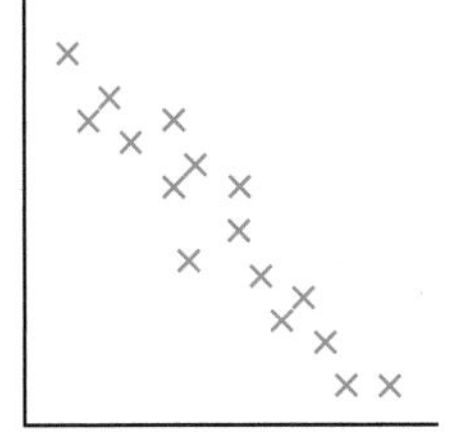

In negative correlation an increase in one set of variables results in a decrease in the other set of variables

Zero or no correlation

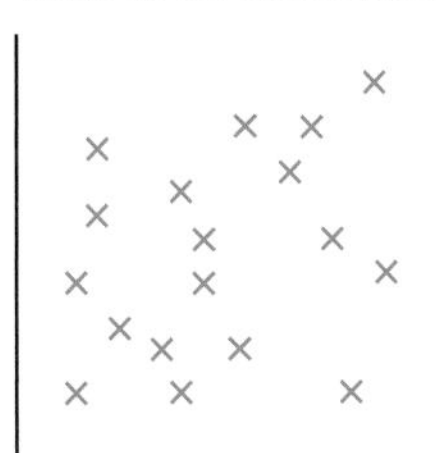

Zero or no correlation is where there is no obvious relationship between the two sets of data

Cumulative frequency diagram – a cumulative frequency diagram can be used to find an estimate for the mean and quartiles of a set of data; find the cumulative frequency by adding the frequencies in turn to give a 'running total'

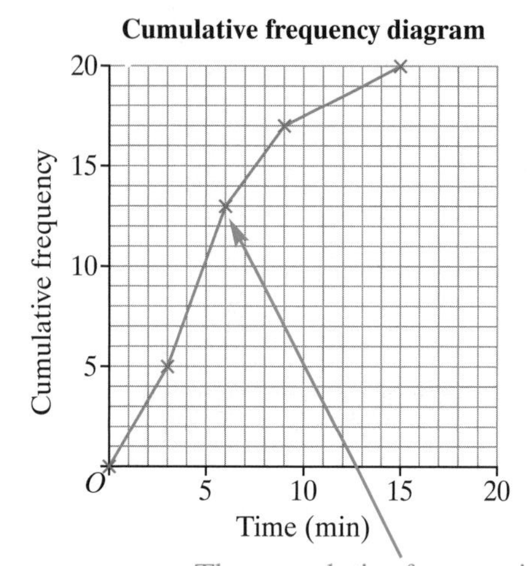

Data – information that has been collected

Data collection sheets – these are used to record the responses to the different questions on a questionnaire; they can also be used with computers to load data onto a database

Dependent events – events are dependent when the outcome of one affects the outcome of the other, for example, taking two successive balls from a box without replacing the first one

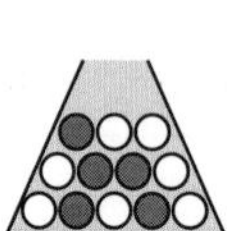
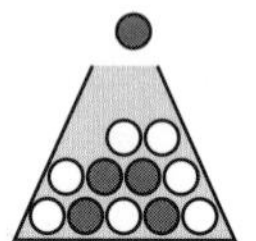
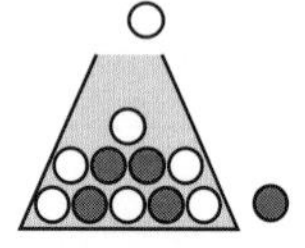

Direct observation – collecting data first-hand, for example, counting cars at a motorway junction or observing someone shopping

Discrete data – data that can only be counted and take certain values, for example, number of cars (you can have 3 cars or 4 cars but nothing in between, so $3\frac{1}{2}$ cars is not possible)

Evens – probability $\frac{1}{2}$, for example, there is an even chance of getting a head or a tail when you toss a coin

Experimental probability or **relative frequency** – this is found by experiment, for example, if you get 6 heads and 4 tails, the experimental probability would be 0.6 for getting a head

Frequency density – in a histogram, the area of the bars represents the frequency and the height represents the frequency density

$$\text{Frequency density (bar height)} = \frac{\text{frequency}}{\text{class width}}$$

Frequency diagram – a frequency diagram is similar to a bar chart except that it is used for continuous data. In this case, there are usually no gaps between the bars

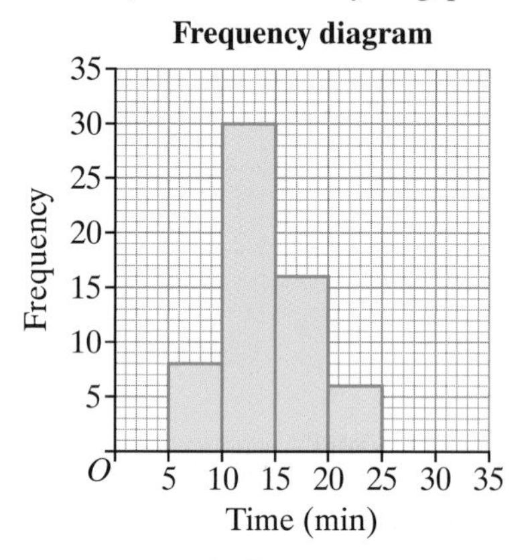

Frequency polygon – this is drawn from a histogram (or bar chart) by joining the midpoints of the tops of the bars with straight lines to form a polygon

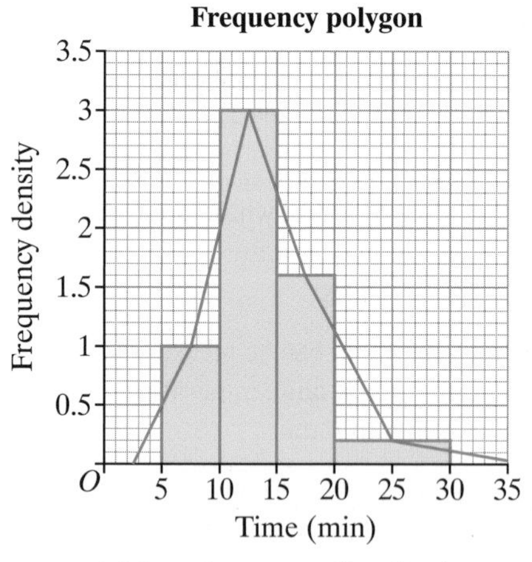

Frequency table or **frequency distribution** – a table showing how many times each quantity occurs, for example:

Number in family	2	3	4	5	6	7	8
Frequency	2	3	8	4	2	0	1

Grouped data – data that has been grouped into specific intervals

Histogram – a histogram is similar to a bar chart except that the *area* of the bar represents the frequency

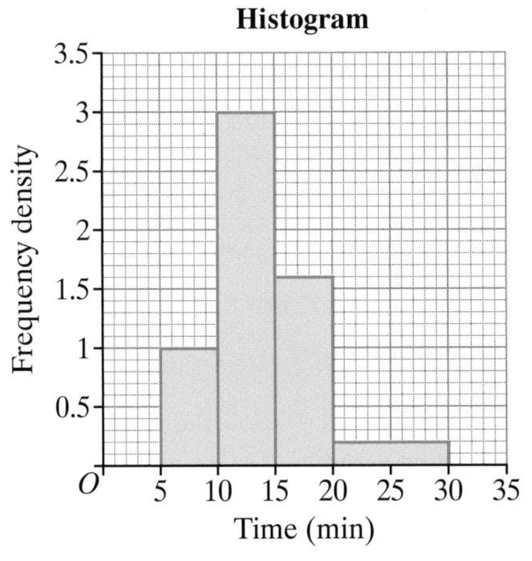

Impossible – an event with a probability 0, for example, the sun turning green

Independent events – events are independent when the outcome of one does not affect the outcome of the other, for example, tossing a coin and drawing a card from a pack

Interquartile range – the difference between the upper quartile and the lower quartile

Interquartile range = upper quartile – lower quartile

Likely – an event with a probability greater than $\frac{1}{2}$, for example, rain falling in November in the UK

Line graph – a line graph is a series of points joined with straight lines

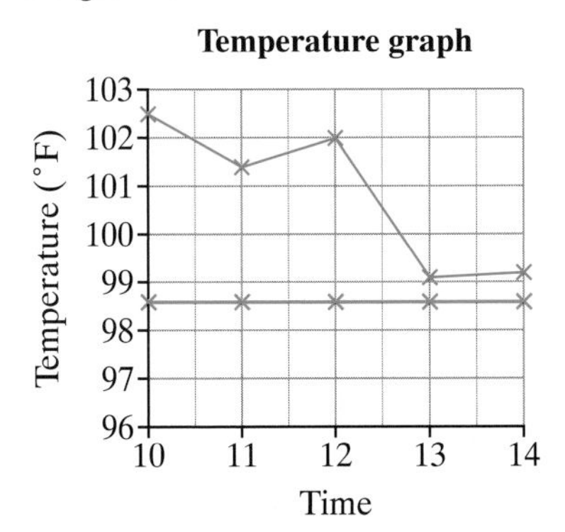

Line of best fit – a line drawn to represent the relationship between two sets of data. Ideally it should only be drawn where the correlation is strong, for example,

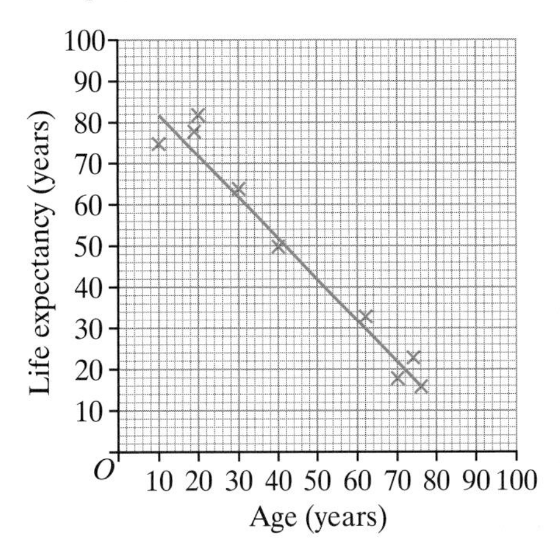

Lower quartile – the value 25% of the way through the data

Mean – found by calculating $\frac{\text{the total of all the values}}{\text{the number of values}}$

Median – the middle value when all the values have been arranged in order of size; for an even set of numbers, the median is the mean of the two middle values

Modal class – the class with the highest frequency

Mode – the value that occurs most often

Moving average – used to smooth out the fluctuations in a time series, for example, a four-point moving average is found by averaging successive groups of four readings

The four-point moving averages can be plotted on the graph as shown

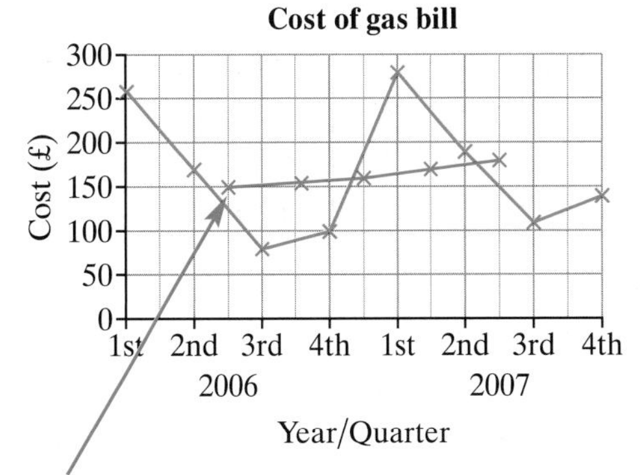

The first four-point moving average is plotted in the 'middle' of the first four points, and so on

Mutually exclusive events – these cannot both happen in the same experiment, for example, getting a head and a tail on one toss of a coin

Opportunity sampling – see convenience sampling

Ordered stem-and-leaf diagram – a stem-and-leaf diagram where the data is placed in order

Number of minutes to complete a task

Stem (tens)	Leaf (units)
1	1 6 7 8 9
2	2 2 7 7 7 8 9
3	1 4 6

Key: 3|4 represents 34 minutes

Outcome – the result of an experiment, for example, when you toss a coin, the outcome is a head or a tail

Outlier – a value that does not fit the general trend, for example,

Pilot survey – a small-scale survey to check for any unforeseen problems with the main survey

Primary data – data that you collect yourself; this is new data and is usually gathered for the purpose of a task or project (including GCSE coursework)

Probability – a value between 0 and 1 (which can be expressed as a fraction, decimal or percentage) that gives the likelihood of an event

Qualitative or **categorical data** – data that cannot be measured using numbers, for example, type of pet, car colour, taste, people's opinions/feelings, etc.

Quantitative data – data that can be counted or measured using numbers, for example, number of pets, height, weight, temperature, age, shoe size etc.

Quota sampling – this method involves choosing a sample with certain characteristics, for example, select 20 adult men, 20 adult women, 10 teenage girls and 10 teenage boys to conduct a survey about shopping habits

Random – a choice made when all outcomes are equally likely, for example, picking a raffle ticket from a box with your eyes shut

Random sampling – this requires each member of the population to be assigned a number; the numbers are then chosen at random

Range – a measure of spread found by calculating the difference between the largest and smallest values in the data, for example, the range of 1, 2, 3, 4, 5 is $5 - 1 = 4$

Relative frequency – see experimental probability

Respondent – the person who answers the questionnaire

Sample space diagrams – see two-way table

Scatter graph – a graph used to show the relationship between two sets of variables, for example, temperature and ice cream sales

Temperature against ice cream sales

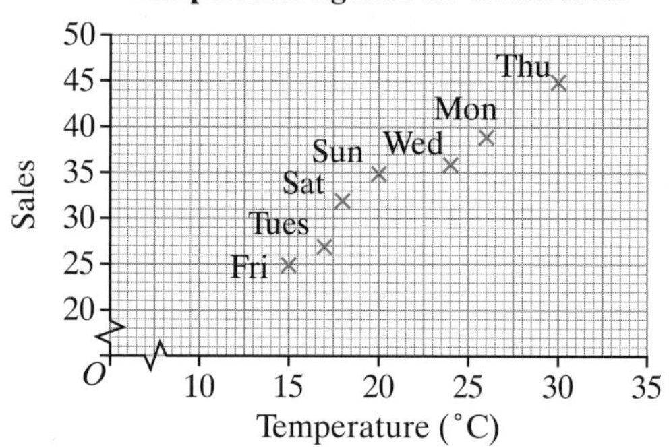

Secondary data – data that someone else has collected; this might include data in books, newspapers, magazines, etc. or data that has been loaded onto a database

Stem-and-leaf diagram – a way of arranging data using a key to explain the 'stem' and 'leaf' so that 3|4 represents 34

Number of minutes to complete a task

Stem (tens)	Leaf (units)
1	6 8 1 9 7
2	7 8 2 7 7 2 9
3	4 1 6

Key: 3|4 represents 34 minutes

Stratified sampling – this involves dividing the population into a series of groups or 'strata' and ensuring that the sample is representative of the population as a whole, for example, if the population has twice as many boys as girls, then the sample should have twice as many boys as girls

Survey – a way of collecting data; there are a variety of ways of doing this, including face-to-face, or via telephone, e-mail or post using questionnaires

Systematic sampling – this is similar to random sampling except that it involves every nth member of the population; the number n is chosen by dividing the population size by the sample size

Tally chart – a useful way to organise the raw data; the chart can be used to answer questions about the data, for example,

Number of pets	Tally
0	𝍸 IIII
1	𝍸 𝍸 II
2	𝍸 II
3	III
4	II

The tallies are grouped into five so that

IIII = 4
𝍸 = 5
𝍸 I = 6

This makes the tallies easier to read

Theoretical probability – probability based on equally likely outcomes, for example, it suggests you will get 5 heads and 5 tails if you toss a coin 10 times

Time series – a graph of data recorded at regular intervals

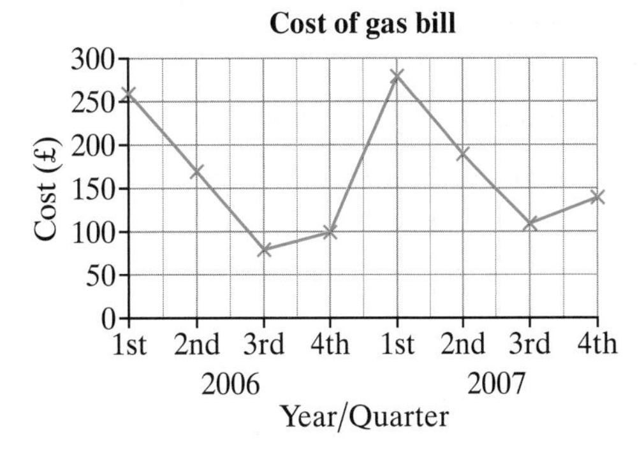

Two-way table or **sample space diagram** – table used to show all the possible outcomes of an experiment, for example, all the outcomes of tossing a coin and throwing a dice

		Dice					
		1	2	3	4	5	6
Coin	Head	H1	H2	H3	H4	H5	H6
	Tail	T1	T2	T3	T4	T5	T6

Unlikely – an event with a probability less than $\frac{1}{2}$, for example, snow falling in August in the UK

Upper quartile – the value 75% of the way through the data